REAL ESTATE AND GIS

Real Estate and GIS focuses on the application of geographic information systems (GIS) and mapping technologies in the expanding property and real estate discipline. Whilst a thorough understanding of location is understood to be fundamental to the property discipline, real estate professionals and students have yet to harness the full potential of spatial analysis and mapping in their work. This book demonstrates the crucial role that technological advances can play in collecting, organising and analysing large volumes of real estate data in order to improve decision-making.

International case studies, chapter summaries and discussion questions make this book the perfect textbook for property and applied GIS courses. Property and real estate professionals including surveyors, valuers, property developers, urban economists and financial analysts will also find this book an invaluable guide to the understanding and application of GIS technology within a real estate industry context.

Richard Reed is Professor and Chair of Property and Real Estate at Deakin University, Australia. He is a registered property valuer and is also the editor of the *International Journal of Housing Markets and Analysis* and author of the best-selling textbook *Property Development*, now in its sixth edition.

Chris Pettit is Professor and Chair of Urban Science at the University of New South Wales, Australia. He is on the board of directors for the CUPUM (Computers in Urban Planning and Urban Management) conference, a board member of the International 'Geo for All' initiative, Co-Chair of the Research Data Alliance (RDA) International Working Group on Urban Quality of Life Indicators and Co-Chair of the International Society for Photogrammetry and Remote Sensing (ISPRS) Working Group on Geographical Visualization and Virtual Reality.

REAL ESTATE AND GIS

The Application of Mapping Technologies

Edited by Richard Reed and Chris Pettit

Routledge
Taylor & Francis Group

LONDON AND NEW YORK

First published 2019
by Routledge
2 Park Square, Milton Park, Abingdon, Oxon OX14 4RN

and by Routledge
711 Third Avenue, New York, NY 10017

Routledge is an imprint of the Taylor and Francis Group, an informa business

British Library Cataloguing-in-Publication Data
A catalogue record for this book is available from the British Library

Library of Congress Cataloging-in-Publication Data
Names: Reed, R. G. (Richard G.), editor. | Pettit, Christopher, editor.
Title: Real estate and GIS : the application of mapping technologies / edited by R.G. Reed & C.J. Pettit.
Description: Abingdon, Oxon ; New York, NY : Routledge, 2018. | Includes bibliographical references and index.
Identifiers: LCCN 2017058005 | ISBN 9781138187979 (hardback : alk. paper) | ISBN 9781138187986 (pbk. : alk. paper) | ISBN 9781315642789 (ebook : alk. paper)
Subjects: LCSH: Real estate business–Geographic information systems. | Real property–Geographic information systems.
Classification: LCC HD1375 .R3725 2018 | DDC 333.330285–dc23
LC record available at https://lccn.loc.gov/2017058005

ISBN: 978-1-138-18797-9 (hbk)
ISBN: 978-1-138-18798-6 (pbk)
ISBN: 978-1-315-64278-9 (ebk)

Typeset in Bembo
by Out of House Publishing

CONTENTS

CONTRIBUTORS

Aida E. Afrooz is a Research Associate at CityFutures research centre, Faculty of the Built Environment at UNSW, Sydney, Australia. She received her Bachelor degree in Urban Development from Azad University of Mashhad, Iran (2006), Master's degree in Urban Design from Universiti Teknologi Malaysia, Malaysia (2009) and Ph.D in Built Environment from UNSW, Australia (2016). She has worked as urban planner and urban designer in private consultancies and local government in Australia and overseas. Her research interests include wayfinding, visual memory and cognitive configurations of the built environment, eye tracking study, 3D modelling and visualisation. She also combines research activities with lecturing, tutoring and thesis supervision.

Robert Buckmaster is principal and founder of the location consultancy *Choice Location Strategists*. He is a property market and investment analyst with Australia-wide property assessment and acquisition experience gained at leading international and national asset advisory firms and institutional investors including Jones Lang LaSalle, Standard & Poor's and AMP. He has advised clients including Bunnings, Investa, Stockland and Moorabbin Airports Corporation. A qualified town planner, valuer and financial analyst his specialisations include trade area profiling, development feasibility assessment and economic impact assessment.

Peter Charles is an independent practicing architect and researcher based at Monash University in Melbourne. His current research interests combine digital generative architecture and urbanism to produce flexible "Porous Fields" that respond dynamically to change at numerous spatial and temporal scales. He previously worked at Kisho Kurokawa in Tokyo and A.R.M in Melbourne, with experience in numerous countries on large international design projects including the Kartal Pendik competition & St Petersberg FIFA Stadium among others, and smaller projects such as the H House in Port Melbourne. He contributes to design methodology regarding complex conditions and growth in Melbourne.

Laura Crommelin is a Research Fellow at City Futures Research Centre in the Faculty of Built Environment at UNSW (Sydney) where she works on projects related to housing affordability, high-density living, urban policy-making and urban renewal. She is also a lecturer in urban planning in the Faculty and her research interests include current trends in post-industrial cities including place branding, place-making and DIY urban revitalisation practices. She holds a Ph.D in urban planning from UNSW, an M.Litt in US Studies from the University of Sydney and a BA/LLB (Hons) from the University of Melbourne.

Benoit Gaudou is an Associate Professor in Computer Science, and more specifically in Modelling and Simulation of Complex Systems, at the University Toulouse 1 Capitole, and a Researcher at the Research Institute of Computer Science of Toulouse. He is interested in the modelling and integration of complex agents in simulation and particularly taking into account trust or emotions in their decision-making process and behaviour. In addition he works on models integrating mathematical and agent-based models. He is taking part to the development and documentation of the GAMA platform.

Hoon Han is an Associate Professor and the Director of City Planning Program and Co-convenor of Smart Cities Research Cluster (SCRC) in the Faculty of the Built Environment at UNSW where he has designed courses on urban economics and economic geography. He serves as an Associate Editor of *City, Culture and Society* journal (Elsevier) and an editorial board member for the journals *Spatial Information Research* (Springer) and *International Journal of Knowledge-based Development* (Inderscience). His research work spans a broad range of topics including housing markets, urban and regional planning, digital interaction and smart cities.

Scott N. Lieske is a Lecturer in Geography within the School of Earth and Environmental Sciences at the University of Queensland. His current research interests are in using spatial tools, data and technology for built environment and natural resources policy development and decision-making. Research topics include city analytics; the relationship between urban form, costs of public services, and infrastructure costs; planning support system theory and implementation as well as regional environmental change. Additional areas of expertise include the effective use of geographic visualisation as a communication and decision-support tool.

Simone Zarpelon Leao is a Research Fellow in Urban Modelling and Simulation at the City Futures Research Centre, University of New South Wales. She works on developing knowledge and methodologies to assist in the generation of new or regeneration of existing urban areas with awareness and consideration of the challenges faced by urban and environmental planning, applying digital geographical information and computer simulation and visualisation as core methodologies. Her multidisciplinary research has been reported in numerous international scientific journals papers and books.

Jian Liang is the Lecturer of Property and Real Estate at Deakin University. His Industry background is as property and real estate analyst in global real estate firms such as Savills and CBRE. His research interest includes real estate finance, real estate valuation and housing markets. His research outcome has been appearing in the international refereed academic journals and conferences in the discipline of property and real estate.

Russell Lowe lectures in Architecture and Computational Design at the University of New South Wales. He coordinates the first year architectural design studio, teaches in the Masters of Architecture graduation studio and delivers electives in Advanced Virtual Reality. Russell is a member of CAADRIA (Computer Aided Architectural Design in Asia) and has published on the use of computer gaming technology in architectural education, design and safety. His research focusses on the repurposing of computer gaming technology to engage with uses and concepts outside of the entertainment industry.

Chris Martin is a Research Fellow at the City Futures Research Centre in the Faculty of Built Environment at UNSW. He has extensive experience in housing policy and practice, and has authored the *Tenants' Rights Manual: A practical guide to renting in NSW* (4th edition). Before joining CFRC, he served as as chairperson for Shelter NSW and Senior Policy Officer for the Tenants' Union of NSW. His current research interest is in tenancy law, private rental market, social housing, marginal rental accommodation, housing and taxation, and housing policy history.

Ryan van den Nouwelant has a background in social and design sciences where he studied and worked in south-east Asian natural resource management before becoming a urban planner. After working as a local government strategic planner he joined City Futures Research Centre at UNSW in 2011. His Ph.D investigated community conflict associated with the mixed land uses in and around Kings Cross, Sydney. An early career researcher his other research interests include examining how urban design, renewal and planning practices can improve housing outcomes in different urban contexts.

Pristine Ong is a postgraduate student at the University of New South Wales with professional experience in communications and online production. She has an interest in urban renewal, city equity, cycling planning and urban design, which has led to a semester of exchange at the University of Copenhagen. Her Master's thesis will examine food justice in urban renewal projects. She completed the Bachelor of International Studies at the University of Sydney, with exchange at Leiden University in the Netherlands. She has also written for online magazines and has been published in literary journals.

Chris Pettit is the inaugural Chair of Urban Science at the University of New South Wales. His expertise is in the convergence of the fields of Urban Planning and Geographical Information Sciences (GIS) where he has published more than 150 peer reviewed papers. For the last 20 years he has been undertaking research and development in the use of spatial information and mapping technologies for envisioning *what if?* scenarios across both urban and regional landscapes. His recent interests also span into applications, development and the critique of geographical visualisation tools including, advanced spatial decision-support systems and city dashboards.

Bill Randolph is Professor and Director of the City Futures Research Centre at UNSW, Australia's leading academic urban research group. Bill has 35 years' experience and an international reputation as a researcher on housing and urban policy issues in the academic, government, non-government and private sectors. His current research focuses on housing markets and policy, urban inequality, urban renewal, high density housing and affordable housing. Bill has been researching aspects of multi-unit living for over 30 years, both in the UK and Australia.

Jon Reades received an undergraduate degree in Comparative Literature but a start-up opportunity led him to code and the mining of large data sets. Working on WAP data he became interested in the impact of mobile telecommunications on cities which led to a MPhil/PhD with BT under Prof. Sir Peter Hall and Prof. Michael Batty. He worked on transport research involving London's Oyster Card (including the 2012 Olympics) before helping to set up the Geocomputation pathway at King's College London. He now focuses primarily on open data because of an interest in replication in the context of complex urban systems and evidence-based policy.

Matthew Reed has focussed his research on retail property and is completing a Ph.D at the University of Southern Qld on the location of retail property and data analytics. All of his industry experience has been in the retail property sector including 10 years employed with the Australian department store 'Myer', before moving to Shopping Centre Management and Development. Matthew has insights into many aspects of retail property including customer base analysis, retail leasing and new retail developments. He is currently undertaking research to provide Australian shopping centre owners with improved knowledge about the perceptions of managing customer service delivered by tenants within their assets.

Richard Reed is the Chair of Property and Real Estate at Deakin University. His industry background is as a property valuer in both the private and public sectors. His research has focused on the multi-disciplinary nature of the property market and more specifically on housing markets, demography and the application of GIS. He is also the editor of *The International Journal of Housing Markets and Analysis* and has authored numerous books, book chapters and journal paper to contribute to the international literature.

Carmela Ticzon received her undergraduate degree in Literature from De La Salle University in Manila and a Master of Environmental Management degree from the University of New South Wales. She is interested in the potential applications of GIS and digital story-telling in community engagement and public participatory planning. Currently she is a Technical Assistant working within the City Analytics team at City Futures Research Centre in UNSW.

Laurence Troy is a Research Fellow and Lecturer at the City Futures Research Centre in UNSW. His research interests lie in the intersection of housing markets, urban governance, urban renewal and social-economic outcomes in Australian cities. His recent work has focused on various facets of urban renewal in the context of a market economy and understanding the economies of developing new community housing across Australian cities. Part of this work involves extensive modelling of renewal feasibility and mapping the dynamics of housing tenure, housing markets and population.

Elizabeth A. Wentz is Dean of Social Science in the College of Liberal Arts and Sciences, Associate Director for the Institute of Social Science Research, and Professor in the School of Geographical Sciences and Urban Planning at Arizona State University. Her research focuses on the development of geographic technologies designed to better understand the urban environment. In 2015-16 she served as President of the University Consortium for Geographic Information Science. She earned her PhD in Geography from the Pennsylvania State University, her MA in Geography from The Ohio State University and her BS in Mathematics from The Ohio State University.

FOREWORD

I am pleased to provide this Foreword for *Real Estate and GIS*, as this is an area of increasing importance in the real estate space, particularly with a stronger research focus now on spatial analysis and real estate. The use of GIS in the real estate area has received increasing coverage over the last 25 years. This has expanded recently with improvements in GIS technologies and a fuller understanding of how GIS can be used effectively as a key tool in understanding the spatial dimensions of real estate. These developments come in perfect timing with the publication of this book and its increasing practical relevance in urban decision-making.

I particularly liked this work's way of addressing the big current issues in real estate going forward, and how GIS can add to our understanding of these issues. These include disruptive technologies, big data, value capture and big city issues, all of which have critical spatial dimensions. The use of case studies in areas such as value capture in transport infrastructure and housing intensity add further to the relevance of the nexus between real estate and GIS. The use of a range of chapter authors enables the real estate and GIS issues to be looked at through different eyes, for a fuller understanding of all the dimensions of the use of GIS in real estate space.

Location is a key aspect of real estate; linking in GIS for a fuller understanding of the spatial aspects of location is a critical issue for practitioners, academics, researchers and students to understand. This book addresses this issue very effectively in an easy-to-read style, with excellent case studies.

I hope you enjoy this book on real estate and GIS, and how GIS can be used effectively in addressing many of the spatial analysis questions around the effective use of real estate.

Professor Graeme Newell
Professor of Property Investment, Western Sydney University, Australia

1

INTRODUCTION

Richard Reed and Chris Pettit

Introduction

As the technological age continues to rapidly evolve and present increasing opportunities to operate more efficiently, this research book addresses an emerging gap relating to the application of geographic information systems (GIS) and mapping technologies to the expanding global property and real estate field. A core strength of this book is the multidisciplinary approach, which provides a unique and innovative investigation into the application of GIS concepts in the real estate sector. The contributors to the chapters in this book are recognised experts in their fields, with each chapter based upon a solid theoretical foundation incorporating research question/s and a case study approach.

This book covers both the GIS and property/real estate disciplines where the focus is placed on the synergy and interaction between both areas, rather than each field in isolation. It has been an increasing trend in many research institutions and industry bodies to collaborate at the discipline level. Therefore this book has been written to assist property stakeholders, GIS stakeholders and those who have an interest in GIS and/or property-related research and application. It successfully strikes a balance between these two fields. Even though there is an expanding body of knowledge about each individual area, there is a slow uptake with the application of GIS in the property and real estate discipline (Sosnowska *et al.* 2016). The real estate discipline itself, especially when referring to the assessment of real estate value by property valuers and surveyors, is largely premised on parameters associated with the *location* of the property (Reed *et al.* 2014). At the same time, surprisingly, there has been relatively little attention given to the underlying fundamentals of *locational intelligence* which is made possible through the use of spatial information, analysis and mapping methods (Michael *et al.* 2013). Although the property and real estate industry has a strong reliance on maps, often they are produced at a very basic level where the user has relatively little knowledge about how to create, manipulate and produce maps, or how to utilise more advanced spatial analysis tools and techniques (Hadiguna *et al.* 2014). Accordingly, this book has been written to fill this gap in knowledge.

The underlying framework for the research in this book is the importance of the *situs* or location of every parcel of land, either in its natural state or one that has been improved in some form or another (Reed 2015). Due to the heterogeneous nature of property, every individual parcel of land has a unique location identifier and no two parcels on earth are identical. This is in contrast to other asset classes, including structural improvements to land (i.e. two buildings can be identical and constructed at the same time), cash and shares/stocks. The scenario is further complicated by the large number of varying land uses observed in cities subject to the requirements of local planning legislation; for example, refer to Appendix B for an extensive list of property classification codes highlighting the large number of alternative land uses. Taken together, these aspects ensure a thorough knowledge of GIS is therefore essential to fully understand how the complex property and real estate market operates individually and collectively, let alone with the addition of other dimensions (e.g. three dimensions) and also the added consideration of change over time (Mishra and Nagar 2009). While other disciplines have made sound advances and have embraced technology in the 21st century, there is now an urgent need for the real estate discipline to catch up with the rapid pace of change as observed in the discipline of GIScience (Krause *et al.* 2012).

The largest single asset class in most countries is property and real estate in its many varying land uses, including areas covered with water, air rights and property rights below the earth's surface; however, the different values (e.g. monetary, social, historical values) of individual and collective real estate is largely of interest to stakeholders, including owners, buyers, lenders, insurers and government agencies (Reed *et al.* 2014). The operation of the real estate market is also of interest in broader society as it affects most people, at the very least via the provision of shelter in the form of residential property. At the same time, there is a strong argument that as the international property discipline and its associated valuation process have been transforming from an 'art' to a 'science', there is a need for additional information about how to use GIS and mapping technologies to better undertake property and real estate analysis. From an international perspective it is common practice to refer to the built environment as 'property' in certain countries (e.g. the UK) and as 'real estate' in other countries (e.g. the US), hence the reference to 'property and real estate'. For the remainder of this text the terms 'property' and 'real estate' are used interchangeably. From a mapping perspective the term 'geographical information systems' is referred to as 'GIS', with the underpinning discipline known as GIScience.

The contributions from this book to the real estate and GIS disciplines

The catalyst for this book is to address the clear gap in knowledge between these two disciplines, as observed by the authors in both the research/university sectors and in industry. Accordingly, this book is focused on research combining real estate and GIS, rather than examining the independent and separate disciplines of real estate and also GIS. As a starting point, there is an assumed basic level of knowledge about property/real estate which will assist in understanding how property is enhanced by mapping applications. For example, the importance of location for a property is commonly accepted (Reed 2015). If the reader is completely unfamiliar with the basic concepts of property prior to commencing this text (e.g. they may come from a pure GIS background), it would be beneficial to become familiar with basic property and real estate principles. In addition, it would be advantageous for each reader to investigate the circumstances surrounding information in their own local area/region, as it is not realistic to cover all contingencies and references to local jurisdictions in this book.

Another important consideration is the availability of real estate and spatial/mapping-related information in a specific area or identified areas of focus. Some of the accepted differences between jurisdictions are listed below, although it is strongly recommended to make individual investigations for individual circumstances:

(a) Detailed property-related characteristics/attributes may be freely available, while certain other information (e.g. owners' names) may be withheld due to privacy law restrictions. The minimum accessible information typically refers to the registered address of the property, the location based on the street address, the land area and other relevant attributes.

(b) Information about improvements (e.g. type of structure, timing/type of renovations), if any, may also be available.

(c) Information relating to other characteristics about each parcel, if applicable (e.g. shape and topography of the land, contamination considerations, potential water views).

(d) Availability of mapping data in softcopy format including GIS shapefiles and other file formats.

(e) Cost of access to property and mapping-related data. In some countries this information is freely available (i.e. zero cost), while in other countries accessing large volumes of detailed data can be practically cost-prohibitive.

(f) Means of accessing data. The simplest approach is to access the data via an online portal over the internet, although alternatives include using a CD (compact disc), a spreadsheet or database package, or possibly via hard copy only.

This research book has been written primarily to provide insights into the array of spatial methods available with applicability for all international property markets. However, each property market will have individual characteristics that need to be fully understood when considering real estate and GIS applications. The main characteristics of real estate can be separated into two core attributes:

• type of land use – e.g. housing, industrial, office, retail, rural property (Appendix B); and
• actual location of the land – e.g. city centre, suburb or rural.

This research book is strong evidence of the rapid growth in research into real estate and GIS over the past 50 years (Krause *et al.* 2012). The catalyst for this book was to highlight the latest up-to-date research and provide a framework for the reader to increase their knowledge and skills about the relevance and application of real estate and mapping. For readers from a real estate background, it may be their first foray into GIS and hopefully will prove an inspiration to embrace advanced mapping programs, rather than relying solely on the simple generic and basic maps that are freely available as web mapping or mobile phone applications. Since the majority of real estate information has some aspect of spatial dimension, GIS outputs can be tailor-made to suit most situations, and an improved understanding of GIS and a property's location directly results in better knowledge about a specific property. At the very least, the mapping can enhance the decision-support process. When relying on real estate as an input, a GIS analyst can embrace an optimal platform to undertake research, including sourcing expansive detailed data and associated analysis, which has 'real life' relevance and provides meaningful results. Most importantly, this research area is evolving as a rapidly emerging area where GIS has an important contribution to make over the long term.

Research in the real estate and property market

The property and real estate discipline is arguably one of the broadest of all research areas (Li *et al.* 2016). Rather than being easily defined and quantified in a similar manner to other disciplines (e.g. architecture, engineering, construction), real estate draws upon of many sub-disciplines which collectively form the property discipline. Examples include:

- Built environment: including, construction, sustainability.
- Finance: including valuation, cost of capital.
- Geography: including urban geography, demography, locational theory.
- Law: including legal ownership, transfer of ownership.
- Engineering.

Each of these disciplines, from a real estate perspective, publish their research in targeted journals in each area. Underlining the relatively new status of the combined real estate and GIS field, there are no journals which are specifically focused on both areas, but rather on either (a) specific aspects of the real estate discipline (e.g. *International Journal of Housing Markets and Analysis*) or aspects for the GIS discipline (e.g. *Computers, Environment and Urban Systems* or *Applied Spatial Analysis and Policy*).

'Art versus science' debate

From a real estate perspective, most of the discussion about the property and real estate market relates to the value of a parcel of land. There are many other discussions and related studies into real estate, such as for statutory planning purposes, but the most common reference to land is linked to the value of the land parcel as an asset at a particular point in time. The underlying premise is that practically all freehold land parcels at one point in time will have their owner-ship transferred; most often this transfer will conform to standard economic theory where the new owner will pay the arm's-length market value in the form of monetary compensation to the seller (Reed 2015). In other words, it is the 'value' of the parcel which forms the reference for much of the discussion in real estate markets, where the location of the property most often adversely affects the property's value (Bowcock 2015).

The accepted approaches to the valuation process are closely aligned to GIS principles and will be substantially enhanced by an improved understanding of mapping fundamentals (Sosnowska *et al.* 2016). To appreciate even a basic understanding of valuation it is essential to consider briefly the background to the valuation discipline, which has often been referred to as the 'art vs. science' debate. Since a valuation is based on a hypothetical sale of a property between a willing buyer and a willing seller, a rudimentary valuation approach is to con-sider the perception of both parties in the market-place based on judgement; this is generally referred to as the 'judgement' component, which actually lacked a substantial scientific basis. However, over time and with the advent of computer technology, there has been a gradual albeit slow uptake of the 'science' component in the process of assessing value. Accordingly, the use of GIS is a substantial component of this modern scientific approach (Krause *et al.*

2012), and whilst it is accepted that many real estate-related activities will include both 'art' and 'science' as per conventional expectations, the emphasis in this book is placed on mapping as a scientific approach that will assist stakeholders in the real estate market as a form of decision-support system.

Technology and real estate

Broadly speaking, the real estate discipline has been relatively slow to adapt to change with reference to technology (Black *et al.* 2003). There are many reasons for this lagged approach, many of which are associated with the fundamental characteristics of the real estate market including:

- the lack of a central trading place or market;
- the unique or homogeneous nature of every property parcel;
- a relatively large number of potential buyers and potential sellers; and
- the lack of reliable and freely available information in the market.

Although other competing asset classes have embraced technology, such as online trading in the share or stock market, many aspects of the real estate market have remained relatively unchanged. However, since the early 21st century there has been a limited uptake of computer technology in the industry by property professionals, for example, assisted by the portability of laptop computers in the field and access to centrally located databases. In many countries most of the maps being referenced remained paper-based up until the mid-1990s, although this has gradually changed to online mapping, as access costs have decreased and the availability of online mapping products has increased. The emerging challenge has been to the ability of real estate stakeholders in using this new technology to improve their efficiency levels (Roig-Tierno *et al.* 2013).

There has remained a sustained level of resistance to the uptake of technology in the built environment, which was acknowledged in the early 1970s (Lee 1973). Arguably, this may be partly due to the long-standing nature of real estate theory which has changed relatively little over time. In other words, the market process for a land transfer is still extremely basic and many stakeholders in the property market have the option of operating at a basic transaction or market analysis level. This means there have not been the same pressures placed on many real estate organisations to increase the level of technology, since many of their day-to-day tasks involve real estate inspections and communication with clients. This 'lagging' behaviour with property experts has been observed at times in other areas; for example, when valuers were confirmed as distinct laggards in a global study of property professionals (Dixon *et al.* 2008).

Real estate and decision-support systems

The real estate industry and its many facets are commonly referred to as a 'people business' and all property stakeholders have continuous on-going interactions with other stakeholders, including clients and government agencies, even if with the intent of sourcing data (Lindh and Malmberg 2008). With the rapid uptake of computer technology in the broader economy, and its substitution for many existing roles in different professions, a similar scenario will arguably not be prevalent in the property industry due to the high proportion of human contact required in property decision-making and in the valuation process, for example. Thus, many aspects of the property market are human orientated and cannot be automatically replaced using computer technology. However, there is significant potential for decision-support systems to assist real estate stakeholders; for example, digital tools such as GIS, dashboards and sensors to assist those working in the property market sector. Accordingly, this is precisely how GIS should be viewed, as a decision-support system to provide the broad real estate discipline with an improved understanding of the spatial and temporal dimensions of the property market and to assist property stakeholders making informed decisions about property. In other words, GIS has the potential to ensure there would be fewer 'unknown' variables and more 'known' variables, which effectively reduces the level of risk in decision-making.

There may be a counter argument that for some property professionals, GIS and mapping may be viewed as non-essential in their day-to-day tasks. This argument may be based on the perceived complexities of GIS and the availability of spatial mapping and modelling tools commonly referred to as 'black box' programs. However, there are two strong arguments to support improved knowledge about GIS as a decision-support system for real estate, as discussed below.

(a) The advent of the technological age has affected most facets of business and daily life, although property and real estate is an area which has embraced computer technology to varying levels. Some property professionals would perceive themselves as 'early adopters', whereas others are simply followers, because they have no alternative choice – i.e. they must avoid a loss situation. Since the underlying property fundamentals are largely based on the importance of locational attributes associated with a land parcel, the task of understanding and quantifying these attributes in a format including enhanced mapping display characteristics will provide a clear competitive advantage.

(b) There are risks associated with making business details largely based on technology which is not fully understood by the user. For those familiar with the somewhat complex discounted cashflow (DCF) valuation approach, it is essential for the user to have a thorough understanding of how the DCF works, rather than simply entering data into a generic DCF application or program. The same approach arguably needs to apply to GIS usage. Therefore, it is essential to understand how the data is structured within a GIS and the array of spatial analysis methods and visualisation techniques now available to provide an in-depth understanding of locational considerations in real estate markets.

International research

The property and real estate market is globally connected in a similar manner to other asset markets (Davis and Zhu 2011). This presents substantial opportunities when property is viewed as a holistic market-place without boundaries, although from a GIS perspective there are data challenges which in turn have affected the level of research conducted in different regions. Some key challenges relate to the availability and quality of data suitable for analysis and visualisation; in addition, a knowledge gap exists in understanding the potential for GIS and location-based intelligence to facilitate a more evidence-based approach to property market analysis, which in turn impacts on the planning and design of urban and rural areas. Hence GIS and associated analytical and visualisation techniques offer the real estate industry with an opportunity to make (a) better economic decisions and (b) better social and environmental decisions to ensure rural and urban areas remain productive, sustainable, liveable and resilient.

Outline of this book

The balance of chapters in this research book is structured as follows.

Chapter 2 – Residential intensification and housing demand: A case study of Sydney, Brisbane and Melbourne

This chapter examines housing intensification across three cities in Australia: Sydney, Melbourne and Brisbane. Specifically, two research questions were addressed: (i) *What are the characteristics of the demand drivers for high-rise living for Sydney, Melbourne and Brisbane?* (ii) *How are the different groups accommodated by this market distributed across the three cities?*

Using Census data collected through the Australian Bureau of Statistics (ABS), the research examined the key components that drive housing intensification across these three rapidly urbanising cites. Choropleth mapping and Principal Components Analysis (PCA) are the key methods used in this research to understand drivers of housing intensification with the ABS Statistical Area level 2 geography. This chapter also introduces an open data platform known as CityData (https://citydata.be.unsw.edu.au/), where the housing intensification maps are stored so others can discover and reuse these map layers. CityData comprises a number of city datasets including housing data, mobility data, city indicator data and much more. Open data platforms are part of the global smart cities agenda and provide the fuel to support city analytics and therefore a deeper understanding of property markets in those cities and jurisdictions that are supporting an open data smart cities agenda (Pettit *et al.* 2018).

Chapter 3 – How disruptive technology is impacting the housing and property markets: An examination of Airbnb

This chapter provides an international case study of three global cities in the context of the shared economy giant Airbnb. Using data on Airbnb made available from AirDNA, this research examines the spatial distribution of Airbnb across Sydney, London and Phoenix. The primary aim of this research was to explore how spatial data analysis can help support cities in responding to the rise of Airbnb in an equitable and efficient way. Therefore, a number of GIS and spatial analysis techniques have been applied in this research, including: dot density mapping, buffering, location quotient analysis and hot spot analysis using Moran's I. Through a data driven approach, this research explores the spatial dimensions of the Airbnb phenomena across these three cities and discusses some of the implications for housing affordability, and other possible impacts.

Chapter 4 – The contribution of GIS to understanding retail property

This chapter investigates the potential of GIS applications to assist in identifying the optimal location and operation of retail destinations, including enclosed retail shopping centres. Owners of retail destinations continually take exhaustive steps to maximise both the asset value and consumer attraction of their centres, however GIS has the potential to enhance the spatial mapping and associated analysis of consumer demand. The research question for this chapter is: *What are the drivers linked to retail customers visiting a specific retail destination rather than a competing destination?* The value of a retail destination is strongly correlated with the centre's net operating income; therefore the turnover and sales achieved by a retail destination are directly accountable to location and ability to attract customers to visit and purchase. An added complication affecting the level of consumer spend in retail destinations is the increased retail competition from online retailers. An examination is conducted of locational drivers of retail destinations, followed by a discussion about consumer behaviour with reference to retail destinations. A case study is based on a hypothetical new store opening for a retail destination where the main drivers were identified as proximity to city centre, location of competing centres and demographic profile of the potential customer base.

Chapter 5 – Modelling value uplift on future transport infrastructure

This chapter presents a prototype spatial decision-support system which has been developed to support city-planners and policy-makers in understanding the potential land value uplift to property prices driven through new transport infrastructure. The key research question explored in this chapter is *Can a rapid spatial analysis decision-support platform be created to support land-use planners' exploration of value uplift scenarios in real-time?* To address this question, the Rapid Analytics Interactive Scenario Explorer (RAISE) prototype toolkit has been developed and tested in Western Sydney for the City of Parramatta. The RAISE toolkit enables city-planners to 'drag and drop' new metro stations into place, and then it rapidly calculates the expected property valuation increase attributable to the proximity to this new proposed infrastructure. The toolkit is fuelled by fine-scale city data at the property level, which includes both physical and neighbourhood characteristics. Underpinning the toolkit is a hedonic price model which determines how much each data point contributes to the overall property value, and presented in this chapter is the first iteration of the toolkit. Future work will see the RAISE toolkit supported by an ensemble of property models over an extended geography. The RAISE toolkit exemplifies what is possible in real-time decision-support as powered through geographical information and modelling.

Chapter 6 – Commercial office property and spatial analysis

This chapter examines the contribution of spatial characteristics including location when constructing a price index to analyse transactions of office buildings. Although the reference is to commercial office buildings which predominantly have an 'office' land use, the application of this methodology is relevant to a diverse range of office buildings. In the analysis a case study approach is used based on transactions of office property buildings located in the Melbourne central business district (CBD) between 2000 and 2015. The research question for this chapter is: *To what extent does spatial dependency exist in the transactional market for office buildings?*

An investigation is undertaken into office building indexes and the spatial dependency of office property. The methodology for the case study is based on analysing every sale of whole or entire office buildings between 2000 and 2015, which equates to 289 detailed CBD transactions.

Chapter 7 – An agent-based model for high-density urban redevelopment under varied market and planning contexts

This chapter presents a novel agent-based modelling approach for evaluating the economic feasibility of residential development proposals. The agent-based model (ABM) has been developed and evaluated in the context of a neighbourhood in the City of Randwick, Sydney. In the context of property and real estate, the specific case study addresses three questions: (i) *What insights can the agent-based model of urban redevelopment provide to users?* (ii) *Can the model be used to investigate potential future scenarios based on neighbourhood design proposals?* (iii) *What are the main benefits and limitations of agent-based modelling for urban redevelopment related research?* The ABM has been built on the GAMA platform which is powered by GIS data and has a 3D visualisation interface for users to view the results. As many cities continue to grow rapidly, there is a need for such tools to support the evaluation of residential development proposals and to understand the economic feasibility. At the same time, such tools need to provide feedback if the proposed urban redevelopment satisfies local planning instruments.

Chapter 8 – Architecture, GIS and mapping

This chapter examines the extent to which GIS and mapping can be applied in the broader context of built environment research with relevance to both architecture and property disciplines. It examines three existing methodologies in relation to the growth and change of urban systems in contemporary cities, then it investigates if they provide an adequate framework and also whether they could be woven together to provide a more robust model relevant to architecture. The chapter then examines methodologies to explore the potential role of voids in an urban area as a result of growth and change in more complex conditions. Owners of retail destinations continually take exhaustive steps to maximise both asset value and consumer attraction of their centres, however GIS also has the potential to enhance the mapping of consumer demand. For this chapter the research question is as follows: *To what extent can GIS and mapping enhance the application of architecture in an urban context?* This chapter utilises case studies to examine the practical implications with the application of GIS in an urban environment such as a city centre. The applied examples of (a) observation, (b) simulation and (c) speculation highlight the three methods of operating on urban conditions when focused on voids.

Chapter 9 – 3D and virtual reality for supporting redevelopment assessment

This chapter examines the utility of three-dimensional (3D) and virtual reality (VR) for supporting the development application (DA) assessment process. Specifically, the research examines the current state of play in DA assessment which typically is supported using two-dimensional (2D) mapping tools. The research then evaluates the potential of 3D and VR technology in the DA workflow. There are two research questions addressed in this chapter: (i) *What are the benefits of utilising 3D modelling in local government sectors for DA assessments?* (ii) *Which level of details (LODs) in a city model can assist planners in assessing DAs?* A number of city-planners from across three municipalities in Sydney are interviewed and involved in a user study. GIS data and 3D building models have been imported into the Unreal game engine platform and participants used an Oculus Rift VR headset to explore a virtual city environment. Users provided feedback on their perception of different 3D models visualised through different levels of detail (LODs). The participants' experience of the virtual environment was also evaluated.

Conclusions

This chapter has highlighted the contribution of the book to the multidisciplinary fields of real estate/property and GIS, with the emphasis placed on the synergy between the two disciplines. The remaining chapters in the book conduct cutting-edge research in a cross-section of different real estate land uses, including housing, share accommodation (e.g. Airbnb), retail and commercial office property. A strength of this book is the diversity of authors who have contributed to the chapters. With the exception of the introduction and the conclusion, each chapter has been authored by a different combination of researchers. The benefits of this approach include the wide range of research methodologies used across the book, as well as the varying approaches to displaying and discussing the information via figures and text. For example, most often the maps would be presented in full colour format but have here been converted to greyscale to facilitate publication; this process itself emphasised the ability to simplify the presentation of maps and figures.

The sequence of the following chapters is not presented in their order of importance, rather it commences with a relatively straightforward approach relating to housing land use (Chapter 2) through to the forward-looking approach of three-dimensional visualisation (Chapter 9). Reference throughout the book should also be made to the supporting information in the appendices; for example, there are land classification codes in Appendix B which highlight the large number of different land uses. The authors of each chapter are experts in their respective fields and their contributions to this book ensure the high level of relevance to the real estate and GIS disciplines.

References

Black, R., Brown, G., Diaz J., Gibler, K. and Grissom, T. (2003). Behavioral research in real estate: a search for the boundaries. *Journal of Real Estate Practice and Education*, 6(1), 85–112.

Bowcock, P. (2015). A discussion paper on valuations for mortgage and the level of house prices. *International Journal of Housing Markets and Analysis*, 8, 27–35.

Davis, E.P. and Zhu, H. (2011). Bank lending and commercial property cycles: Some cross-country evidence. *Journal of International Money and Finance*, 30, 1–21.

Dixon, T., Cololantonio, A., Shiers, D., Reed, R., Wilkinson, S. and Gallimore, P. (2008). A green profession? A global survey of RICS members and their engagement with the sustainability agenda. *Journal of Property Investment and Finance*, 26(6), 460–481.

Dolan, E.L., Elliott, S.L., Henderson, C., Curran-Everett, D., St. John, K. and Ortiz, P.A. (2017). Evaluating discipline-based education research for promotion and tenure. *Innovative Higher Education*, 43, 31–39.

Hadiguna, R.A., Kamil, I., Delati, A. and Reed, R. (2014). Implementing a web-based decision support system for disaster logistics: A case study of an evacuation location assessment for Indonesia. *International Journal of Disaster Risk Reduction*, 9, 38–47.

Krause, A.L. and Bitter, C. (2012). Spatial econometrics, land values and sustainability: Trends in real estate valuation research. *Cities*, 29(2), S18–S25.

Lee Jr, D.B. (1973). Requiem for large-scale models. *Journal of the American Institute of Planners*, 39(3), 163–178.

Li, R.Y.M. and Chau, K.W. (2016). *Econometric Analyses of International Housing Markets*. Abingdon: Routledge Studies in International Real Estate.

Lindh, T. and Malmberg, B. (2008). Demography and housing demand – what can we learn from residential construction data? *Journal of Population Economics*, 21, 521–539.

Michael, K. and Clarke, R. (2013). Location and tracking of mobile devices: Uberveillance stalks the streets. *Computer Law and Security Review*, 29(3), 216–213.

Mishra, S. and Nagar, R. (2009). GIS in Indian retail industry – a strategic tool. *International Journal of Marketing Studies*, 1, 50–57.

Pettit, C., Bakelmun, A., Lieske, S.N., Glackin, S., Thomson, G., Shearer, H., Dia, H. and Newman, P. (2018). Planning support systems for smart cities. *City, Culture and Society*, 12, 13-24.

Reed, R.G. (ed.) 2015. *The Valuation of Real Estate*. Canberra: Australia.

Reed, R.G. and Sims, S. (2014). *Property Development*, 6th edition. Abingdon: Taylor & Francis.

Roig-Tierno, N., Baviera-Puig, A. and Buitrago-Vera, J. (2013). Business opportunities analysis using GIS: the retail distribution sector. *Global Business Perspective*, 1, 226–238.

Sosnowska, M. and Karszinia, I. (2016). Methodology for mapping the average transaction prices of residential premises using GIS. *Polish Cartographic Review*, 48(4), 161–171.

2

RESIDENTIAL INTENSIFICATION AND HOUSING DEMAND

A case study of Sydney, Brisbane and Melbourne

Bill Randolph, Aida E. Afrooz and Chris Pettit

Introduction

Australian cities, in common with many other cities worldwide, are undergoing a major process of densification with the development of high-rise multi-unit housing. While in many emerging economies these new high-density residential areas are built on otherwise undeveloped land, in advanced economies densification has often involved a process of renewal and redevelopment, often of older redundant or lower value commercial and industrial sites. The transition to higher density urban renewal in Australian cities has had a transformative impact over the past two decades (Rosewall and Shoory 2017). Multi-unit development accounted for half of all residential development nationally in 2015–16 compared to just over a quarter in 2009, and a total of 114,000 apartments were approved in Australia in 2016, a 275% increase over 2009 (Housing Industry Association 2017). This fundamental change in land use outcomes has been fostered by the pursuit of planning policies that have promoted 'compact city' mixed-use, higher density urban renewal and infill development as the principal mechanisms for accommodating expected population growth (Forster 2006; OECD 2012). Higher density residential development, often focused on urban renewal schemes around 'magnet' infrastructure, has become a dominant feature in location decisions about new housing supply in cities (Searle and Filion 2010). As a result, the physical character of major cities, especially in inner and middle suburban town centres, is changing rapidly while significant claims about the benefits of urban compactness, notably improved urban amenity, affordability, accessibility, diversity, liveability, productivity and sustainability, now dominate the rhetoric of planning and development discussion both in Australia and internationally (Searle and Filion 2010). This is despite these benefits remaining a contested area of scholarship (Neuman 2005).

In Australian cities this process has been most pronounced in the inner areas and along key transport corridors and suburban centres (Adams 2009; Curtis 2012). Indeed, metropolitan planning strategies have increasingly been focused on directing such redevelopment into these areas for at least the last two decades. What is less clear, however, is who exactly is being housed in this new high-density sector. With a predominance of one- and two-bedroom apartments, it is clear that the new higher density market is not primarily targeting families.

While this may reflect the declining importance of families in the composition of households in Australia in recent years, they nevertheless remain an important component of the population. On the other hand, the growth in older households, particularly older single people, has prompted planners to argue the new apartment market is a suitable response to meet their needs. But while the higher density market could be seen to be catering for this demand, so far there is little evidence that the older population has downsized into this form of accommodation to any great degree (Judd *et al.* 2014). Finally, the predominance of investors in the apartment market, with over half of all apartments being rented out (Randolph 2006), means that the market is unlikely to accommodate households in the established stages of their life cycle, where home ownership is the preferred and predominant tenure choice. There is little hard evidence that the established pattern of homeownership as the aspirational choice of established households has faltered, or that this group is moving into smaller rental apartments out of choice. The analysis and methodology presented in this chapter aims to develop a better understanding of the structure of the demand for the apartment market in Australia's three largest cities using recent Census data to identify the various demand sub-markets that underpin its growth. The findings should assist planners to develop better land use polices that take account of who is actually being housed through this rapidly growing higher density market that their plans are strongly promoting.

In particular the analysis draws on two key insights developed in earlier versions of this methodology. The first is that one of the distinctive features of the higher density housing market is its spatial fragmentation, with pockets of apartments located in areas that have been zoned for this kind of dwelling, while it is largely missing from other areas of the city where low-density prevails. A second characteristic which makes for difficulties in analysing the sector is that it is a three-dimensional (3D) market, often with a mix of demand groups within a single building, let alone within a neighbourhood. This makes the traditional two-dimensional (2D) spatial analysis approaches to unpacking and mapping the sub-structure characteristics of higher density housing increasingly problematic. The approach adopted here brings these two insights together by operationalising the concept of *spatially discontinuous housing markets* (Randolph 1991) which proposes that, in reality, housing markets are differentiated across space in a series of interpenetrating and spatially overlapping sub-segments in which similar locations or indeed buildings cater for different mixes of these segments. The method also uses GIS to show the spatial interpenetration of these sub-markets within the higher density population which assists in dealing with the long-standing 'ecological fallacy' problem which bedevilled the simple 2D spatial analyses common in earlier approaches (Robinson 1950; Openshaw 1984).

Research aim and questions

This chapter elucidates the higher density housing demand characteristics of the three major capital cities in Australia, namely Sydney, Melbourne and Brisbane. It compares the socio-economic characteristics of residents of apartment dwellings in the above-mentioned three metropolitan regions using Principal Component Analysis (PCA). The study therefore aims to identify the socio-economic profiles of the different household groupings that have been accommodated by the intensification of residential land use through apartment development in Australian cites. The approach develops the methodology used by Randolph and Tice (2013) to explore the sub-market structure of the Sydney and Melbourne apartment markets which used 2006 Census data. This research updates the analysis to 2011 data, extends the number of variables used to widen the scope of the analysis and includes Brisbane. The research sought to address two questions:

> Research question 1: *What are the characteristics of the demand drivers for high-rise living for Sydney, Melbourne and Brisbane?*
> Research question 2: *How are the different groups accommodated by this market distributed across the three cities?*

Background and literature review

It has generally been accepted that housing markets are differentiated across space. The origins of such understanding date back to the earliest days of housing studies. Charles Booth's *Descriptive Maps of London Poverty, 1898–99* was one of the first attempts to map social divisions in a major urban area, identifying the class structure of London down to the street scale. The impact of the Chicago school of urbanists in the 1920s and 30s led to further theoretical and empirical developments (Park and Burgess 1925), while the post-war emergence of social area analysis took the analysis of urban social residential structure to yet another level of sophistication and argued that urban residential structure was formed around three broad factors: economic status, family status and ethnicity (Timms 1971). This led the way to the first computer-aided analysis of the socio-spatial structure of cities in the 1960s, using new techniques such as factor analysis to test these theories, aided by the wider availability of small area census data. Similarly, urban economists have also sought to better understand the sub-structure of urban housing in terms of environmental characteristics (Galster 1987), property type (Dale-Johnson 1982) and the role of real estate professionals (Hamilton 1998). A more sociological perspective led to the housing class model developed by Rex and Moore (1967), which proposed a socio-economic hierarchy of housing sub-markets that would have distinctive spatial correlates (note: for a summary and critique of this tradition see Basset and Short 1980).

All of these approaches basically assumed a 2D low-density city in which sub-markets were essentially more or less homogenous in terms of broad characteristics. However, these assumptions no longer are relevant in cities that are undergoing widespread physical renewal, processes of both gentrification and the emergence of new concentrations of urban disadvantage, changing housing market dynamics, as well as increasingly being built in three dimensions through higher density development. It can be argued, therefore, that cities have become increasingly spatially fragmented and disrupted such that simple 2D patterns of socio-spatial structure have reduced utility. At the same time, the traditional methods of analysis and mapping these patterns of socio-spatial difference are also becoming increasingly redundant. An alternative approach to understanding how housing sub-markets are located in space is needed.

The concept of *spatially discontinuous* housing sub-markets offers a way to conceive geographically interpenetrating sub-markets occupying similar neighbourhoods. Randolph (1991) suggested the concept of *spatial discontinuity* provides a more realistic approach to characterise the geographic expression of high-density housing markets, which are more properly understood as vertically integrated and overlapping in spatial terms, rather than spatially homogenous areas of similar housing types and household characteristics:

The housing market in any one locality is therefore characterized by a hierarchy of overlapping socially and spatially defined segments … Each segment will have its own distinctive economic, social and locational characteristics, defined both in terms of the structure of provision on which it is based and the social characteristics of the population it accommodates

(Randolph 1991, p.34).

The 'structure of provision' refers to the institutional attributes (e.g. rental or owned; strata titled; social or private sector) and the physical form of the dwelling. This approach explicitly recognises the overlapping of different sub-markets in space and is therefore particularly valuable in the case of the higher density market, where blocks of apartments on a single site may accommodate different housing sub-markets *within* the same building, such as renters and owners, or students and young families. The method developed to undertake the analysis explicitly accommodates this characteristic of the higher density housing market using a multi-factor approach.

Housing intensification

Housing intensification in this context refers to the relatively recent development of higher density housing in the form of multi-unit dwellings. Promoted by planning authorities over the last two decades as a solution to urban sprawl, and enthusiastically adopted by the development industry, the move to higher density multi-unit development has been a significant new feature of many inner and middle city areas, especially where it is associated with planned urban renewal. Figure 2.1 illustrates the rapid escalation in the development of multi-unit housing in Australia over the last decade. For the first recorded time it was noted that apartment buildings exceeded single houses by mid-2016, although the levels of development have fallen back somewhat since. Nevertheless, the high-rise multi-unit revolution has been a major feature of housing markets in Australia's major cities and is unlikely to retreat significantly, especially given predicted population growth pressures on those cities. Several characteristics of this new market are worth mentioning. First, they are overwhelmingly built

FIGURE 2.1 Building Approvals, Australia 2006 to 2017
Source: Housing Industry Association, Media Release 30 August 2017.

TABLE 2.1 The number of apartments in each case study city in 2001 and 2011

City	Number of apartments				2001–2011	
	2001	%*	2011	%*	No	%
Sydney	343,518	23.9	391,887	25.8	48,369	14.1
Melbourne	177,579	14.3	219,111	15.3	41,532	23.4
Brisbane	69,886	11.6	85,751	11.7	15,865	22.7
Australia	923,139	13.1	1,056,237	13.6	133,098	14.4

*percentage of occupied private dwellings of each city in regional level (e.g. 23.9% of occupied private dwellings in Greater Sydney were flats). Note: metropolitan area is the statistical level of this table. Source: Australian Bureau of Statistics (2001, 2011a).

as one- or two-room dwellings. Second, they are largely targeted at and sold to investors who will rent them out. Third, they have been concentrated in central and inner-city areas as well as along transport corridors, existing town centres and in major master-planned renewal precincts on land owned by government or redundant transport or industrial sites. The latter, in particular, are characterised by high-rise apartment buildings of up to 20 storeys or more in some central business districts. This is a relatively new feature of housing in Australian cities. These characteristics provide an immediate market structure largely determining the range of households that is typically accommodated. The analysis in this chapter seeks to reveal the household characteristics of the underlying sub-markets that have been accommodated in this emerging higher density market for the three cities.

In Australia, while multi-units have been a feature of some suburbs in Sydney and, to a lesser extent, Melbourne for many years (Butler-Bowden and Pickett 2007; Lewis 2000), more recently there has been a marked shift in the scale of development of this kind of housing. In Brisbane multi-unit development was not prominent until the last decade or so. It is therefore much less developed than in the other two cities. Table 2.1 summarises the growth of the multi-unit apartment market in the three case study cities in the decade between 2001 and 2011. The table highlights the much larger established apartment sector in Sydney, where apartments accounted for a quarter of the housing stock in 2011, compared to 15% in Melbourne and 12% in Brisbane, but with Melbourne and Brisbane experiencing larger proportional growth over the decade. The much smaller scale of the apartment market in Brisbane is also transparent. These three cites accounted for 66% of the total 1,056,237 apartments in Australia in 2011: 37% in Sydney, 21% in Melbourne and 8% in Brisbane (Australian Bureau of Statistics 2001, 2011a).

Research methods

The three largest Australian cities, being Sydney, Melbourne and Brisbane, as defined by the ABS's Urban Centre and Localities (UCL) classification of urban areas with populations of one million or more from the 2011 Australian Census, were chosen as the spatial focus of the analysis. Building upon the methodology developed by Randolph and Tice (2013), small area census tracts (i.e. Statistical Area 1 – typically around 200 households) characterised by higher density dwellings were selected for the analysis. Data relating to 48 socio-demographic variables for households classified as living in flats or apartments in buildings of three or more storeys were extracted from the ABS Census Table Builder tool for these SA1 tracts (Australian

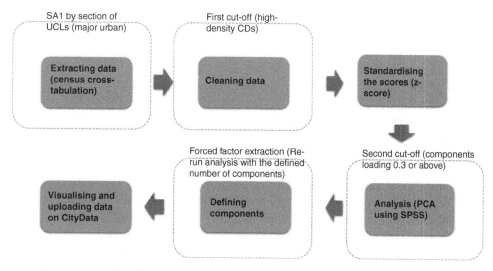

FIGURE 2.2 Methodological framework

Bureau of Statistics 2011b). With this approach a comprehensive profile of households living in multi-unit housing in SA1s in the three cities was obtained.

Data was then prepared and a first 'cut-off' identified, as SA1s with 30 or more apartments were selected to exclude low-density Census tracts from the analysis. Therefore the bulk of the higher density population was included. Next, data was normalised using z-scores to enable authors to compare scores from different normal distributions. The standardised data was used as the input for the next stage of the analysis, which involved running a Principal Component Analysis (PCA) to analyse the datasets in these higher density SA1s to explore the underlying socio-demographic structures of the households who live in this form of housing. The approach used in PCA is described in the following section.

Once the PCA had been undertaken, a second 'cut-off' identified components deemed to be significant in explaining the majority of the variance in the dataset. After defining the meaningful components, forced factor extractions (Varimax rotations) were run for these defined components. Once the model was run for each city, the results were mapped using the loadings for each Census Statistical Area 1 on each of the resulting components. This process is illustrated in Figure 2.2. The last phase involved the spatial visualisation of the results and data storage using the City Futures' CityData platform. CityData, which is powered by 'GeoNode' being an open source Geospatial content management system, is a web-based application and platform for sharing, storing and managing city data – see https://citydata.be.unsw.edu.au/. CityData comprises various open spatial data layers on housing, mobility and healthy city indicators, which can be accessed programmatically by online mapping portals such as the Australian Urban Research Infrastructure Network (AURIN) (Pettit *et al.* 2015) or downloaded and used in a desktop GIS package or another analytics software package (Leao *et al.* 2017).

Principal Components Analysis (PCA)

Principal Component Analysis (PCA) is a multi-variate method '...to take p variables $X_1, X_2,..., X_p$ and find combination of these to produce indices $Z_1, Z_2,...,Z_p$ that are uncorrelated...'

(Manly and Alberto 2016, p.103). PCA is a method of extracting the most important variables, in the form of component, from a large set of variables (Analytics Vidhya Content Team 2016). In essence it is a 'variable-reduction' technique, which reduces a larger set of variables into a smaller set that captures most of the variance in the original variables (Laerd Statistics 2015). In other words, PCA creates a new measurement scale which identifies those variables in a dataset which have the highest explanatory value. A number of statistical validity checks, known as assumptions, were run to ensure the dataset was suitable for PCA. The first check relates to linearity (i.e. correlation) between all variables using a correlation matrix, because the best results for PCA can be obtained when the variables are correlated either positively or negatively (Manly and Alberto 2016). All variables had at least one correlation coefficient greater than 0.3 resulting from a correlation matrix undertaken prior to the procedure being run. The second check was sampling adequacy. Two tests were undertaken for this purpose including the Kaiser-Meyer-Olkin (KMO) measure of sampling adequacy for the overall data set and Bartlett's test of sphericity (Laerd Statistics 2015). KMO values can range from 0 to 1 with values above 0.6 suggested as a minimum requirement for sampling adequacy and values less than 0.5 considered as 'unacceptable' for running PCA (Kaiser 1974). KMO tests were above 0.6 for all the three cases in this study. Bartlett's test of sphericity also indicated these data were suitable for PCA with a $P<0.05$.

With PCA the first resulting 'component' accounts for the greatest amount of total variance within the dataset, with subsequent components explaining successively smaller amounts. PCA expresses this in a summary statistic for each component identified by the procedure, called an 'eigenvalue'. As the analysis generates as many components as there are original variables, choosing the number of components that can be taken to offer the greatest explanatory power is a subjective judgement. However, four major criteria can be used to guide this decision (Laerd Statistics 2015). Subjective assessment plays a part in all these criteria except the first one, which is the component's eigenvalue score.

The second criterion is the proportion of total variance explained by the components, both individually and cumulatively. An individual component should explain at least 5–10% of the total variance and the chosen components should account for at least 60% or 70% of the cumulative variance (Laerd Statistics 2015). The third criterion is the scree plot test which graphically displays the eigenvalue of each component in descending order of magnitude. The components to be retained are those before the significant 'inflection point' in the distribution, where the graph begins to level out and little additional explanation is gained from retaining more components. The final criterion is interpretability. It is based on the concept of 'simple structure' and whether the final Rotated Component Matrix makes intuitive sense, and is typically defined as where each variable loads strongly on only one component and each component loads strongly on at least three variables.

Results

Sydney

In order to determine the number of components to be subjected to spatial analysis, the four criteria mentioned earlier in the previous section were investigated. The PCA results for Sydney revealed nine variables with eigenvalues greater than 1, which explained 28.1%, 14.4%, 12.2%, 7.6%, 5.1%, 3.7%, 2.7%, 2.5% and 2.3% of the total variance, respectively. The percentage of variance suggested the first five components, which accounted for 67.5% of

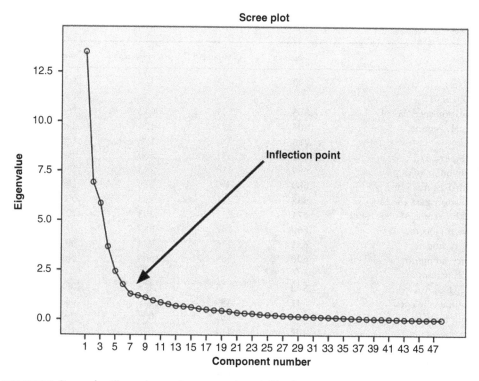

FIGURE 2.3 Scree plot illustrating socio-economic variables for Sydney, SA1, 2011

cumulative variance, were significant. While the scree plot indicated the first seven components could be retained (Figure 2.3), only a four-component solution met the interpretability criterion. Consequently, four components were selected to meet all the above-mentioned criteria, explaining 62.4% of the total variance.

These four components were extracted before running the factor analysis procedure to end up with the rotated component matrix as shown in Table 2.2, which shows the component loadings for the first four factors.

The four housing demand components resulting from the analysis for Sydney can be characterised as follows:

Component 1: Overseas students and Gen Y renters. This group is strongly characterised by young adults (i.e. 15–34 years of age), either tertiary students or not in the labour force who migrated to Australia predominately from North-East, South-East and South Central Asia between 2001 and 2011. The component loads highly on households renting privately, people living in group households, with middle range household incomes (i.e. A$800–1,499 per week). When working, they were in lower skilled service and administrative occupations. Significantly, this component loaded heavily on the apartment variable, indicating a strong correlation with the newer neighbourhoods with concentrations of high rise. This component explained 28% of total variation in the dataset.

Component 2: The economically engaged. This component loaded highly on middle higher income households (i.e. above A$1,400 per week) who are in full-time employment and are engaged in professional and managerial occupations. Residents are either from North-West

TABLE 2.2 Rotated structure matrix with Varimax rotation of a four-component socio-economic variables for Sydney, 2011

Variables	Component 1	Component 2	Component 3	Component 4
Migrants who arrived 2001–2011	**.890**	-.003	.225	.169
Population aged 25–34	**.863**	.344	.042	.167
Privately renting	**.803**	.227	.124	.298
Flats	**.783**	.334	.198	.175
Unemployed	**.762**	.091	.360	.091
Group households	**.689**	.350	.243	.130
North-East Asia born	**.689**	.000	.219	.100
Population aged 15–24	**.685**	.041	.258	.003
Weekly income A$800–1,499	**.671**	.237	.293	.127
No or nil income	**.660**	.032	.047	.076
Tertiary students	**.631**	.116	.044	.084
Service, administrative and sales occupations	**.620**	.569	.308	.069
South-East Asia born	**.590**	.155	.275	.038
Part-time employed	**.541**	.489	.412	.036
Labour force not stated	**.496**	.136	.093	.291
Home buyer	**.492**	.114	.078	.109
Southern Central Asia born	**.460**	.261	.357	.182
Weekly income A$2,500–3,999	.181	**.888**	.160	.124
Managers, professionals and technician occupations	.388	**.879**	.033	.165
Weekly income over A$4,000	.102	**.797**	-.159	-.106
North-West Europe born	.048	**.784**	.319	.166
Full-time employed	.533	**.773**	.099	.147
Family households without children	.523	**.743**	.042	.084
Oceania and Antarctica born	.081	**.736**	.284	.379
Population aged 35–54	.369	**.642**	.480	.090
Home owned outright	.217	**.625**	.283	.312
Weekly income A$1,500–2,499	.561	**.582**	.186	.104
Americas born	.322	**.533**	.169	.105
Population aged 55–64	.022	**.476**	.418	.456
Sub-Saharan Africa born	.051	**.343**	.175	.064
Semi-detached, terrace, and townhouse	-.149	.290	.014	.165
Family households with children	.113	.297	**.874**	.015
Population aged under 15	.134	.131	**.853**	.062
Multiple family households	.192	.226	**.641**	.016
Machine operators and labourer occupations	.456	.306	**.634**	.076
Migrants arrived 1981–2000	.564	.011	**.593**	.146
Separate house	.438	.183	**.576**	.178
North Africa and Middle East born	.225	.308	**.480**	.189
Weekly income under A$400	.212	.228	.030	**.815**
Population aged over 65	.176	.174	.035	**.777**

TABLE 2.2 (Cont.)

Variables	Component 1	Component 2	Component 3	Component 4
Not in labour force	.350	.026	.390	**.696**
Weekly income A$400–799	.395	.153	.330	**.655**
Public renting	.076	.302	.014	**.649**
Lone person households	.390	.367	.312	**.643**
Migrants arrived 1941–1960	.299	.354	.149	**.558**
Migrants arrived 1961–1980	.119	.222	.511	**.529**
Migrants arrived 1895–1940	.147	.180	.112	**.419**
Southern Eastern Europe born	.053	.069	.318	**.415**

Note: major loadings for each factor are in bold. Where a variable loaded highly in more than one component, the highest loading was considered in the results. For example, the second component of 'full-time employed' variable is in bold since it has higher loadings for component 2 rather than component 1.

Europe, the Americas or Australia and New Zealand with some migrants from Sub-Saharan Africa. They are largely a middle-aged group (35–54), with substantial numbers owning their apartment outright and are couple households without children. Component 2 accounted for 14.4% of total variation.

Component 3: Multicultural family households. This group comprises multi-family and family households with children aged below 15 and who are engaged in manual occupations. It includes migrants from North Africa and the Middle East who migrated to Australia between 1981 and 2000. Interestingly, this component was also characterised by a loading on separate houses which indicates concentrations of multi-unit dwellings in otherwise lower density suburban locations. Component 3 accounted for 12.2% of total variation.

Component 4: Lower income retirees. This group is dominated by low-income and aged households. This group loads highly on people over 65 years old who are not in the labour force, are lone-person households and there is a notable loading on public housing. There is a strong loading on migrants from Southern Europe who arrived before 1980. Component 4 accounted for 7.6% of total variation.

Figure 2.4 illustrates the distribution of the loadings on the first four components for Sydney at the SA1 level. Only SA1s with component loadings above 0.3 are presented in these maps, to emphasise the spatial differentiation between the four components.

The map in Figure 2.4.a, for the first component (i.e. overseas students and GenY renters), illustrates the distribution of this group in relation to locations of various universities in Sydney, especially in the central area around University Technology Sydney (UTS) and the University of Sydney, eastern suburbs around University New South Wales (UNSW), North-West Sydney around Western Sydney University (Parramatta campus) and in the north around the Macquarie University. Other linear and scattered patterns can be seen along the metro line from the central business district (CBD) to Parramatta and along main streets in south Sydney which could also indicate where the part-time employed population of this component are living.

The second component (i.e. the economically engaged, Figure 2.4.b) are concentrated mostly in the inner city and eastern beachside suburbs and also along the Parramatta River and the North shoreline. The third component (i.e. multicultural family households, Figure 2.4.c) is more broadly scattered in the middle and western suburbs but includes concentrations

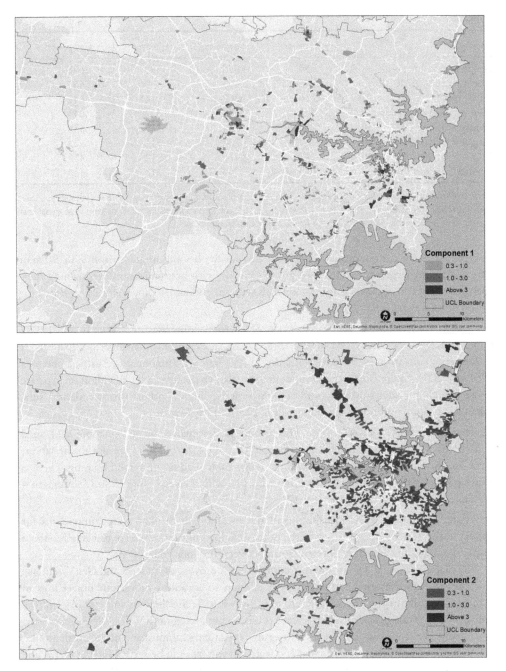

FIGURE 2.4 Location of SA1s with high loadings of the four factors in Sydney, 2011

FIGURE 2.4 (Cont.)

around suburban town centres, with the Canterbury to Bankstown area in the middle western suburbs standing out, as well as along the rail line to the north of Sydney's CBD. Similar to the third component, the fourth component (i.e. lower income retirees, Figure 2.4.d) is scattered away from the CBD, with notable concentrations in the western suburbs as well as towards the south-west, especially associated with suburban public housing estates.

The mapping of the SA1 factor loadings therefore clearly indicates a distinctive geography to these four major demand sub-markets for apartments in Sydney. As can be seen from the four maps, while there are clear differences between the overall distributions of the SA1 loadings for the four components, they also overlap a considerable amount in places. The value of the PCA approach is that it allows the interpenetration of the various sub-markets to be exposed as well as highlighting the 'spatially discontinuous' nature of the higher density housing market in reality.

Melbourne

For the Melbourne analysis the criteria testing resulted in the following outcome. The first assessment generated eight eigenvalues greater than 1, explaining 14.4%, 7.2%, 5.9%, 3.5%, 2.4%, 1.8%, 1.3% and 1.1% of the total variance respectively. The second criteria, assessing the percentage of variance, suggested the first five components could be chosen. These components accounted for 69.6% of the cumulative variance. The third assessment criterion, the scree plot, indicated six components before the inflection point could be retained (see Figure 2.5). The

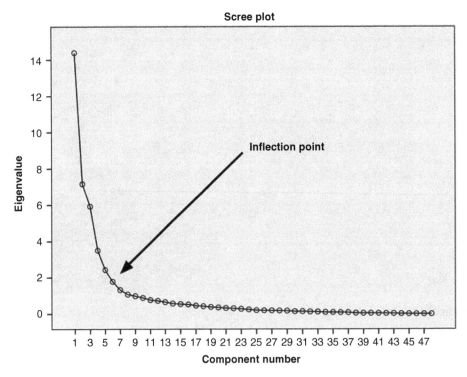

FIGURE 2.5 Scree plot illustrating socio-economic variables for Melbourne, SA1, 2011

fourth component selection method based on interpretability, suggested that retaining the first three components was appropriate. In order to meet all the above-mentioned criteria, the first three components, which together explained 57.2% of the total cumulative variance in the data set, were therefore selected for further analysis.

Next, the forced Varimax rotation was run with the three factors which resulted in the rotated component matrix, as shown in Table 2.3. The results identify the following components for higher density demand in Melbourne and are broadly in line with those for Sydney.

Component 1: Overseas students and Gen Y renters. As in Sydney, this group is dominated by young adult (15–34 years old) migrants from South-East, North-East and Southern Central Asia who arrived in Australia between 2001 and 2011 and are typically students and outside the workforce. Although this component loads heavily on private renters, it also loads on home buyers, indicating inclusion of some economically active households or students whose parents have bought housing for them. Group households as well as lone person households and households with very low incomes also featured prominently. Component 1 explained 30% of total variation for the Melbourne case study.

Component 2: The economically engaged. The main characteristic of this group is that it appears to encompass two sub-groups. The first group comprises full-time employed managerial and professional occupations with high-level household income (i.e. A$2,500 and above). This sub-group assumed to be younger middle aged (35–54) Australian and Oceania and Antarctica couples (i.e. family households without children). The other sub-group includes part-time employees engaged in service and administration occupations with medium levels of income (A$1,500–2,500). This sub-group is associated with migrants from North-West Europe and the Americas as well as areas characterised by semi-detached, terrace and/or townhouses, which in Melbourne are essentially an inner-city built form. Component 2 explained 15% of total variation of the dataset.

Component 3: Multicultural family households. This group is dominated by young children and family households with children with many living in multi-family (i.e. multi-generation) households. Migrants in this group typically arrived in Australia between 1961 and 2000, mostly from Southern Eastern Europe, and a smaller proportion from North Africa, Middle East and Sub-Saharan Africa. This group is engaged in lower skill manual employment occupations (i.e. machine operators and labourers) with lower income (A$400–799 weekly). The majority of this group is not in the labour force. There is also an aged group over 55 in this component who seem to be associated with longer term migration (1941–1960) and out-right home ownership. Overall, component 3 explains 12% of total variation.

The map in Figure 2.6.a for Component 1 (i.e. Overseas students and Gen Y renters) are surprisingly associated with the location of the various universities in Melbourne, especially in the central area as well as suburban concentrations around Monash and La Trobe in the southeast and Victoria University in the west. The second component (i.e. the economically engaged, Figure 2.6.b) are more centrally concentrated around the Melbourne CBD with a buffer distance into the surrounding regions which encompasses Bayside, inner-east Melbourne, and south-eastern Melbourne. In contrast to the second component, the third component (i.e. multicultural family households, Figure 2.6.c) is scattered in middle and outer suburban areas away from the CBD, especially in an arc around the west and north of the city and into the south-east around Dandenong and south-west towards Geelong.

TABLE 2.3 Rotated structure matrix with Varimax rotation of a three-component socio-economic variables for Melbourne, 2011

Variables	Component 1	Component 2	Component 3
Migrants arrived 2001–2011	**.906**	.098	.156
Population aged 15–24	**.845**	.138	.081
Private renting	**.843**	.295	.036
Unemployed	**.827**	.056	.322
Tertiary students	**.796**	.134	.191
South-East Asia born	**.780**	.098	.136
North-East Asia born	**.761**	.026	.071
No or nil income	**.759**	-.063	.156
Flats	**.730**	.254	.246
Group households	**.705**	.395	.299
Population aged 25–34	**.693**	.567	.070
Home buyer	**.606**	.014	.029
Lone person households	**.559**	.321	.054
Labour force not stated	**.516**	.097	.074
Weekly income A$800–1,499	**.491**	.474	.454
Southern Central Asia born	**.458**	.064	.445
Weekly income under A$400	**.421**	.284	.410
Migrants arrived 1895–1940	.033	.013	.006
Managers, professionals and technician occupations	.252	**.936**	.062
Full-time employed	.362	**.878**	.156
Weekly income A$2,500–3,999	.076	**.871**	.179
Family households without children	.343	**.775**	.200
Weekly income A$1,500–2,499	.339	**.769**	.278
Oceania and Antarctica born	.056	**.768**	.513
Weekly income over A$4,000	.079	**.684**	.237
Population aged 35–54	.133	**.678**	.562
Part-time employed	.320	**.665**	.404
North-West Europe born	.055	**.622**	.065
Service, administrative and sales occupations	.407	**.580**	.531
Americas born	.430	**.525**	.020
Semi-detached, terrace and townhouse	.006	**.360**	.015
Population aged under 15	.079	.249	**.780**
Family households with children	.065	.397	**.748**
Separate house	.182	.206	**.739**
Migrants arrived 1961–1980	.107	.120	**.717**
Weekly income A$400–799	.357	.084	**.662**
Machine operators and labourer occupations	.407	.343	**.649**
Not in labour force	.470	.105	**.640**
Multiple family households	.180	.100	**.617**
Southern Eastern Europe born	.001	.066	**.597**
Migrants arrived 1981–2000	.475	.037	**.548**
Population aged 55–64	.137	.420	**.543**
Population aged over 65	.185	.072	**.479**
Migrants arrived 1941–1960	.272	.041	**.477**
Home owned outright	.254	.345	**.468**
North Africa and Middle East born	.379	.187	**.412**
Sub-Saharan Africa born	.343	.112	**.365**
Public renting	.284	.377	.205

Note: major loadings for each factor are in bold. Where a variable loaded highly in more than one component, the highest loading was considered in the results.

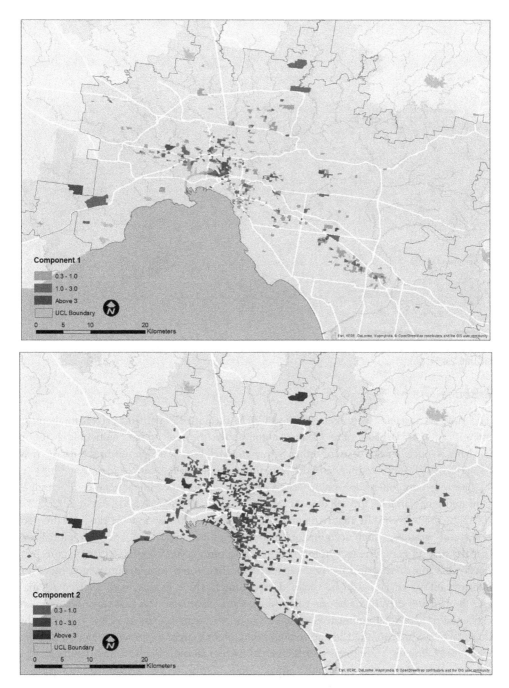

FIGURE 2.6 Location of SA1s with high loadings of the three factors in Melbourne, 2011

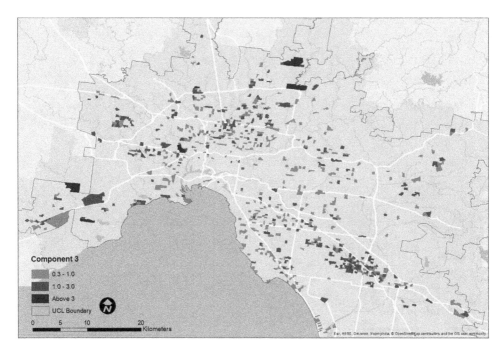

FIGURE 2.6 (Cont.)

Brisbane

The same 48 variables were used for the PCA to measure the demand drivers for housing intensification in Brisbane. However, as noted above, the size of the multi-unit sector in Brisbane is much smaller than for either Sydney or Melbourne, which means the dataset may be subject to greater variability and the resulting components less easy to identify.

The four component assessment criteria produced the following outcome: Eigenvalues indicated nine variables greater than 1: 29.0%, 15.3%, 10.7%, 7.1%, 5.4%, 3.1%, 2.9%, 2.2% and 2.2% respectively. The first five components are within the acceptable range of the percentage of variance, which account for 67.5% of cumulative variance. The scree plot shows that the first six components before the inflection points could be retained (see Figure 2.7). Finally, the fourth selection method suggested the first three components which meet the interpretability criterion can be retained. Accordingly, three factors were selected to analyse the demand drivers for housing intensification of Brisbane. The first three components explained 54% of total cumulative variance. Next, the Varimax extraction was run with the three factors which resulted in the rotated component matrix as shown in Table 2.4, from which the following components for higher density demands in Brisbane are identified:

Component 1: The economically engaged. In the Brisbane case, component 1 showed a much greater variety of resident profile compared to Sydney and Melbourne, and exhibited some of the characteristics of both component 1 and 2 in those other two cities. This group loads heavily on employed persons in middle to higher income white collar employment and childless households. Private rental is the dominant tenure. There are a wide range of adult households, including recent migrants and students as well as employed people. This group includes both full-time and part-time employers and is dominated by renters who are living

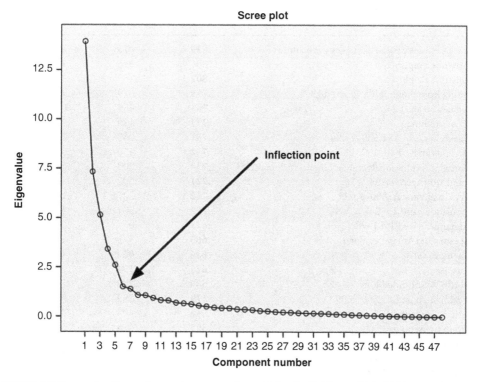

FIGURE 2.7 Scree plot illustrating socio-economic variables for Brisbane, SA1, 2011

in flats. Migrants in this group were either recently migrated to Australia (2001–2011) which might be the case of tertiary students, or have migrated in the last 30 years (1981–2000). Overseas migrants were born in a variety of continents, including Europe and the Americas, to Africa, the Middle East and Asia. For Brisbane, component 1 explained 29% of total dataset variation.

Component 2: Lower income retirees. This group is dominated by adult residents aged over 55, loading heavily on medium density housing forms – semi-detached terrace and townhouses – in outright home ownership and with low weekly income (A$400–800). Workers are engaged in lower skill occupational employments (i.e. machine operators and labourers). However this group is mostly not in the labour force which correlates to the high level of retirees, as well as multi-generational (i.e. multi-family) households with children. Typically, migrants in this group arrived in Australia more than 30 years ago (i.e. between 1941–1980). Component 2 explains 15% of total variation.

Component 3: Low income. This group is characterised by non-working lower income residents (i.e. under A$400 weekly) living in public housing. There is a moderate loading on migrants from South-East Asia. Component 3 explains 10% of total variation.

Figure 2.8 illustrates the distribution of the first three components in Brisbane. As in the other figures, only SA1s with loadings above 0.3 are presented.

The map (see Figure 2.8.a) for the first component (i.e. the economically engaged) illustrates the distribution of this group close to CBD and around the inner city. This ties in closely with the latest wave of higher density development in Brisbane, adjacent to the central

TABLE 2.4 Rotated structure matrix with Varimax rotation of a three-component socio-economic variables for Brisbane, 2011

Variables	Component 1	Component 2	Component 3
Service, administrative and sales occupations	**.839**	.015	.296
Part-time employed	**.806**	.016	.173
Full-time employed	**.802**	.159	.486
Family households without children	**.799**	.079	.258
Population aged 25–34	**.784**	.459	.032
Private renting	**.777**	.110	.267
Weekly income A$1,500–2,499	**.751**	.060	.284
Migrants arrived 2001–2011	**.742**	.450	.344
Managers, professionals and technician occupations	**.734**	.302	.554
Population aged 35–54	**.721**	.302	.437
Weekly income A$800–1,499	**.712**	.227	.126
Population aged 15–24	**.702**	.304	.331
Migrants arrived 1981–2000	**.673**	.078	.005
Oceania and Antarctica born	**.663**	.573	.287
Unemployed	**.634**	.039	.447
Flats	**.625**	.154	.209
North-West Europe born	**.580**	.299	.232
Weekly income A$2,500–3,999	**.579**	.282	.587
Tertiary students	**.558**	.498	.319
Group households	**.558**	.613	.067
Americas born	**.555**	.435	.129
Lone person households	**.512**	.330	.314
North Africa and Middle East born	**.502**	.192	.426
South-East Asia born	**.499**	.457	.315
Sub-Saharan Africa born	**.482**	.014	.026
No or nil income	**.462**	.392	.437
Southern Central Asia born	**.450**	.195	.244
Southern Eastern Europe born	**.425**	.038	.078
Labour force not stated	**.354**	.058	.294
Weekly income over A$4,000	**.334**	.232	.646
Semi-detached, terrace and townhouse	.159	**.704**	.243
Population aged over 65	.056	**.673**	.261
Weekly income A$400–799	.299	**.673**	.454
Migrants arrived 1961–1980	.380	**.636**	.049
Migrants arrived 1941–1960	.163	**.623**	.137
Home owned outright	.316	**.570**	.196
Population aged 55–64	.505	**.553**	.217
Population aged under15	.372	**.550**	.205
Not in labour force	.467	**.527**	.517
Machine operators and labourer occupations	.438	**.472**	.246
Family households with children	.446	**.452**	.419
Multiple family households	.158	**.390**	.050
Weekly income under A$400	.217	.534	**.565**
South-East Asia born	.449	.277	**.480**
Public housing renting	.131	.334	**.479**
Home buyer	.215	.020	.270
Separate house	.262	.317	.049
Migrants arrived 1895–1940	.009	.184	.076

Note: major loadings for each factor are in bold. Where a variable loaded highly in more than one component, the highest loading was considered in the results.

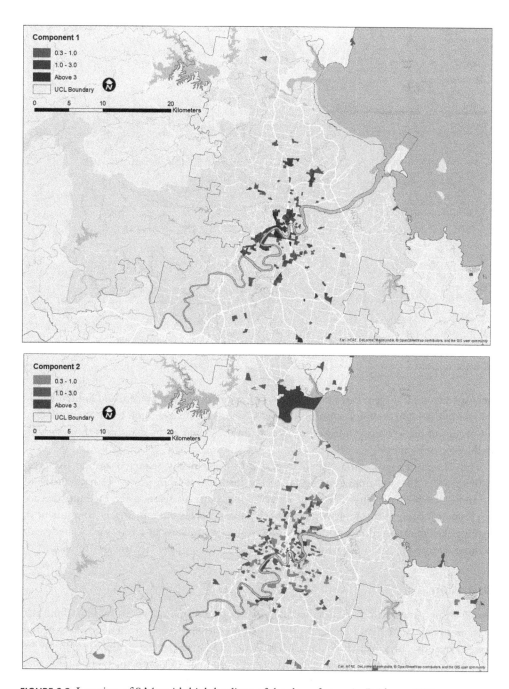

FIGURE 2.8 Location of SA1s with high loadings of the three factors in Brisbane, 2011

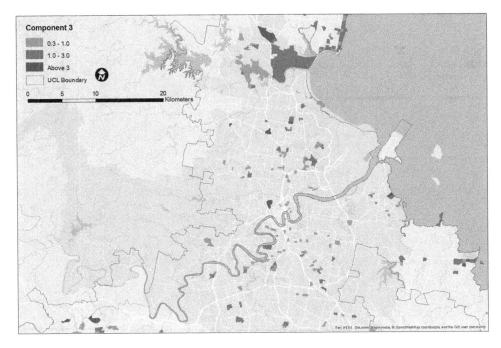

FIGURE 2.8 (Cont.)

inner suburbs such as New Farm on the river. This suggests the more recent and limited development of the higher density sector in Brisbane is not extensive enough for more distinctive sub-groups of the apartment market to become differentiated at this stage as is the case with Melbourne or Sydney.

The second factor (i.e. the retired, Figure 2.8.b) are mostly scattered around the edge of Brisbane's CBD and numerous SA1s located further out from the CBD, particularly in the inner-north of the city. This may reflect this group's preference for medium density developments beyond the high-rise central city market. The third map (i.e. component 3, Figure 2.8.c) illustrates the distribution of low income residents in clusters associated with medium density public housing; for example, in the far west of the city towards Ipswich.

Discussion

The PCA undertaken in this study has addressed both research questions by unpacking the most prominent underlying demand groups that are driving the apartment markets in the three case study cities, namely Sydney, Melbourne and Brisbane, and mapping them out at the SA1 Census tract level. The analysis isolated a number of dominant components that accounted for the majority of variation in the dataset used. These components were broadly comparable in Sydney and Melbourne but were less distinctive in Brisbane. In both Sydney and Melbourne the most prominent component could be characterised as a sub-market that accommodated a high proportion of student and younger renter households, the latter in moderately paid white-collar employment. Migrants from Asia were also strongly represented. These cohorts were concentrated in the CBDs of these two cities but also along rail routes and local clusters. especially around suburban universities. The second most prominent component

TABLE 2.5 Comparison of the proportion of total variation accounted for by the dominant components in the three case study cities

City	Students/Gen Y renters	Economically engaged	Multi-cultural families	Low income retirees	Public renters	Total
Sydney	28%	14%	12%	8%	–	62%
Melbourne	30%	15%	12%	–	–	57%
Brisbane	29%	–	–	15%	10%	54%

was characterised by highly economically active households with middle to higher incomes and with a notable scoring on outright home ownership. This was the upper end of the multi-unit market with a strong locational presence in the central and inner-city and higher amenity locations, such as around Sydney Harbour. A third group was characterised by a predominantly lower income migrant population with multi-generational families and children. Employment was focused on manual occupations. This component scored heavily in more suburban locations well beyond the inner city. However this group was not prominent in Brisbane. Instead the higher density market in Brisbane featured two other groups, being low income retirees and public renters, the latter not prominent in either the Sydney and Melbourne markets. Low income retirees also featured in Sydney, although to a lesser extent and more widely spread across the city compared to Brisbane.

A comparison of the proportion of total variance accounted for by these various components for the three cities as shown in Table 2.5 illustrates how similar the high-density markets in Melbourne and Sydney markets have become, with the first three components representing very similar proportions of the variation in the datasets for these two cities. In contrast, as noted above, in Brisbane the first two components seem to be largely merged together; this is a probable reflection of the much smaller size and more recent development of the higher density sector in this city, where it has had limited scale or time to diversify into more distinctive spatial sub-markets, as has clearly happened in the other two cities.

Conclusion

This chapter presented outcomes from a Principal Components Analysis of the higher density housing markets in Australia's largest cities, Sydney, Melbourne and Brisbane, as based on 2011 Census small area data. The analysis has been able to illustrate clear cleavages in these markets, which are differentiated socially into a limited number of distinctive sub-market components defined by household socio-economic characteristics of residents in higher density housing. In terms of the first of the two research questions posed at the beginning of the chapter, the analysis pointed to a sector driven by five key sub-markets across the three cities: a dominant sub-market characterised by younger renters and students, a second sub-market of middle to higher income economically active households, a third sub-market comprised of lower income multi-cultural families, a fourth sub-market of retirees, often on lower incomes, and a fifth sub-market typified by public renters. But the analysis also showed that the balance of these sub-markets varied between the three cities. While Sydney and Melbourne revealed broadly similar sub-markets, Brisbane's high-density sub-markets were significantly different, with a much less differentiated structure. These differences are related to the different trajectories of the higher density markets in the three cites, with Brisbane's higher density market a much more recent development, with the exception of a small, but nevertheless significant suburban public housing sector.

In relation to the second research question, there were clear differences in the spatial distributions of the five sub-markets, albeit with a variable degree of spatial overlay and inter-penetration. These spatial patterns reflect different trajectories of the higher density market in each city and, again, each city had its own distinctive characteristics. The analysis demonstrated therefore the value of the methodology in unpacking this form of housing market into its main constituent components and illustrating the geographical variations in the distributions of these characteristics.

The chapter therefore demonstrates how the approach can be effectively used to unpack complexity in urban scale analyses using spatially disaggregated datasets of this kind. The latter feature exemplifies the notion of spatially disaggregated and discontinuous housing markets proposed by Randolph (1991). This concept is particularly appropriate for the analysis of higher density markets where traditional two-dimensional spatial analysis is inadequate. Further research could be undertaken to include more explicit indicators of housing sub-market structure, such as property sales and rental data, housing affordability and stress and residential turnover, as well as other information on building typology and age, for example. These data are available at point level from state government records, but adapting this kind of point level data into the analysis presented here would require additional data manipulation and aggregation. The method could also be extended to other spatially disaggregated datasets and urban contexts. The final set of housing intensification spatial data layers for the three cities can be accessed via CityData: https://citydata.be.unsw.edu.au/.

Acknowledgements

The authors would like to acknowledge the Australian Research Council (ARC). Funding through an ARC Linkage Infrastructure, Equipment Facilities (LIEF) grant has made this research possible. The authors would also like to thank Dr Peter Geelan-Small from Stats Central, UNSW, for advice on Principal Components Analysis and to Andrew Tice from the NSW Department of Planning and Environment for advice on how the model is calibrated and presented in spatial output form. Any errors of omission or commission as entirely those of the authors.

References

Adams, R. (2009). Transforming Australian cities for a more financially viable and sustainable future: transportation and urban design. *The Australian Economic Review*, 42(2), 209–216.

Analytics Vidhya Content Team (2016). Practical guide to Principal Component Analysis (PCA) in R and Python. Retrieved from www.analyticsvidhya.com/blog/2016/03/practical-guide-principal-component-analysis-python/.

Australian Bureau of Statistics (2001). Census quickstats: dwelling structure. Retrieved from www.censusdata.abs.gov.au/census_services/getproduct/census/2001/quickstat/105?opendocument.

Australian Bureau of Statistics (2011a). Census quickstats: dwelling structure. Retrieved from www.censusdata.abs.gov.au/census_services/getproduct/census/2011/quickstat/1GSYD?opendocument.

Australian Bureau of Statistics (2011b). Census tablebuilder - saved tables. Retrieved from https://auth.censusdata.abs.gov.au/webapi/jsf/tableView/openTable.xhtml.

Basset, K. and Short, J. (1980). *Housing and Residential Structure: Alternative Approaches*. London: Routledge and Kegan Paul.

Butler-Bowden, C. and Pickett, C. (2007). *Homes in the Sky: Apartment Living in Australia*. Miegunyah Press Melbourne University Publishing.

Curtis, C. (2012). Transitioning to transit-orientated development: The case of Perth, Western Australia. *Urban Policy and Research*, 30(3), 275–292.

Dale-Johnson, D. (1982). An alternative approach to housing market segregation using hedonic pricing data. *Journal of Urban Economics*, 11, 311–332.

Forster, C. (2006). The challenge of change: Australian cities and urban planning in the new millennium. *Geographical Research*, 44(2), 173–180.

Galster, G. (1987). Residential segregation and interracial economic disparities: A simultaneous-equations approach. *Journal of Urban Economics*, 21(1), 22–44.

Hamilton, T. W. (1998). Real estate market segmentation. *Assessment Journal*, 5, 56–69.

Housing Industry Association (2017). Media release, 30 August 2017.

Judd, B., Liu, E., Easthope, H., Davy, L. and Bridge, C. (2014). Downsizing amongst older Australians (No. 214). Australian Housing and Urban Research Institute.

Kaiser, H. F. (1974). An index of factorial simplicity. *Psychometrika*, 39, 32–36.

Laerd Statistics (2015). Principal Components Analysis (PCA) using SPSS Statistics: Statistical tutorials and software guides. Retrieved from https://statistics.laerd.com/spss-tutorials/principal-components-analysis-pca-using-spss-statistics.php.

Leao, S. Z., Lieske, S. N., Conrow, L., Doig, J., Mann, V. and Pettit, C. J. (2017). Building a national-longitudinal geospatial bicycling data collection from crowdsourcing. *Urban Science*, 1(3), 23; doi:10.3390/urbansci1030023.

Lewis, M. (2000). *Suburban Backlash: The Battle for Melbourne, The World's Most Livable City*. Melbourne: Blooming Books.

Manly, B. F. J. and Alberto, J. A. N. (2016). *Multivariate Statistical Methods: A Primer*, 4th edition. CRC Press.

Neuman, M. (2005). The compact city fallacy. *Journal of Planning Education and Research*, 25(1), 11–26; doi.org/10.1177/0739456X04270466.

OECD (2012). *Compact City Policies: A Comparative Assessment*. Paris: OECD.

Openshaw, S. (1984). Ecological fallacies and the analysis of areal census data *Environment and Planning A*, 16(1), 17–31.

Park, R. E. and Burgess, E. W. (1925). *The City*. University of Chicago Press.

Pettit, C. J., Barton, J., Goldie, X., Sinnott, R., Stimson, R. and Kvan, T. (2015). The Australian Urban Intelligence Network supporting Smart Cities, in S. Geertman, J. Stillwell, J. Ferreira and J. Goodspeed (eds.), *Planning Support Systems and Smart Cities, Lecture Notes in Geoinformation and Cartography*, 243–259. Springer.

Randolph, B. (1991). Housing markets, labour markets and discontinuity theory. In J. Allen and C. Hamnett (eds.), *Housing and the Labour Markets. Building The Connections*, 16–51. London: Unwin Hyman.

Randolph, B. (2006). Delivering the compact city in Australia: current trends and future implications. *Urban Policy and Research*, 24(4), 473–490; doi.org/10.1080/08111140601035259.

Randolph, B. and Tice, A. (2013). Who lives in higher density housing? A study of spatially discontinuous housing sub-markets in Sydney and Melbourne. *Urban Studies*, 50(13), 2661–2681; doi.org/10.1177/0042098013477701.

Rex, J. and Moore, R. (1967). *Race, Community and Conflict: A Study of Sparkbrook*. London: Oxford University Press.

Robinson, W. S. (1950). Ecological correlations and the behaviour of individuals. *American Sociological Review*, 15, 351–357.

Rosewall, T. and Shoory, M. (2017) Houses and apartments in Australia. *Reserve Bank of Australia Bulletin, June Quarter*.

Searle, G. and Filion, P. (2010). Planning context and urban intensification outcomes. *Urban Studies*, 48(7), 1419–1438; doi.org/10.1177/0042098010375995.

Timms, D. (1971). *The Urban Mosaic: Towards a Theory of Residential Differentiation*. Cambridge University Press.

3

HOW DISRUPTIVE TECHNOLOGY IS IMPACTING THE HOUSING AND PROPERTY MARKETS

An examination of Airbnb

Chris Pettit, Carmela Ticzon, Jon Reades, Elizabeth A. Wentz, Pristine Ong, Chris Martin, Laurence Troy and Laura Crommelin

Introduction

In an era of smart cities and ubiquitous computing, people are increasingly connected through digital platforms. Castells (2010) labelled this shift the 'information age', which has seen the rise of the 'network society' and a new economy driven by new and disruptive technologies. Danneels (2004, p.247) defines disruptive technologies as ones which fundamentally change the basis of competition, and the importance attached by investors to firms seen 'to disrupt' is evident in their valuations: ride-sharing app Uber valued at US$50 billion (Loizos 2017); banking and finance disruptor SoFi valued at US$4 billion (Wang 2017); and short-term rental platform Airbnb valued at US$31 billion (Lunden 2017). Many of these firms operate via self-organising, peer-to-peer transactions and are collectively referred to as the 'sharing' or 'gig' economy (Fisher 2015; EY 2016). However, the rapid growth of these platforms, and the dominance of a few players in each industry, is increasingly prompting questions about who gets to share in the value these firms create, and who bears the cost.

In this chapter we explore the spatial patterns of activity on one such platform, Airbnb, which has radically reshaped the way many tourists plan and book accommodation. Airbnb has generally argued that it promotes responsible tourism (e.g. AirbnbCitizen.com), generating additional income for governments and residents while not competing with existing downtown hotels and other tourist infrastructure. However, evidence is mounting that Airbnb is disrupting both accommodation and housing markets, resulting in higher rents and the displacement of long-term residents in favour of short-term visitors. Using fine-scale listings data collected by the firm AirDNA in three cities on three continents – Sydney, London and Phoenix – we undertake a spatial analysis to explore whether proximity to the city centre is a key factor for Airbnb listings and also attempt to identify hot spots of activity across these three international cities. This analysis sheds some light on Airbnb's tourist and housing market impacts, in ways that may help these cities to respond to Airbnb equitably and efficiently. More importantly, however, the analysis demonstrates how spatial data analysis tools can support the formulation of city insights, while also highlighting the importance of using these tools alongside detailed, on-the-ground knowledge of the cities and their people.

What is Airbnb and what issues does it raise for cities?

Airbnb is an online platform that enables people (hosts) to rent out residential accommodation – either the whole premises, a private room or a shared room. Hosts and guests can review each other's performance, and this information is collected and displayed by the platform to inform prospective hosts and guests. Created in 2007 as 'Airbed and Breakfast', the company's first listing was the founders' living room in San Francisco during a major design conference. It has now spread worldwide, with Paris leading in Airbnb listings, followed closely by London, New York and Rio de Janeiro (McCarthy 2016). As of March 2017, the company was valued at \$31 billion, and has grown so rapidly that the hotel industry recognises it as the market leader for short-term accommodation (Lane and Woodworth 2016).

What are the benefits of Airbnb?

The appeal of Airbnb for tourists is the possibility of overall cost-savings, amenities (e.g. space, kitchen, laundry), and a more 'authentic' local experience (Guttentag 2013). Airbnb claims to complement existing hotel accommodation, by pointing out that the majority of its listings are outside city centres (Lawler 2012). Alexander (2015) contends that this can erode the traditional spatial divide between the urban tourist and the suburban local, enabling tourists to participate more fully in the life of the city. Airbnb has also argued that, by distributing tourists more widely across cities and regions, the economic benefits of tourism are more widely distributed, with spending by both Airbnb guests and hosts on local products and services producing substantial local economic multiplier effects (Dwoskin 2016).

With regard to hosts, Ellen (2015, p.783) suggests that the success of Airbnb illustrates that there is excess capacity within existing housing stock and that many people are willing to forgo some privacy and share their homes in return for compensation. Jefferson-Jones (2015) has suggested that short-term rentals on Airbnb allows owners to share the financial burden of home ownership, an argument that is also strongly promoted by Airbnb itself (Hunt 2016).

What are the disbenefits of Airbnb?

Contrary to Airbnb's claim that it complements hotel tourism, evidence from that sector indicates that it is being badly impacted. In a study of the Texas hotel market, Zervas, Proserpio and Byers (2016) find that short-term lets through Airbnb have had a negative impact on local hotel revenues, with lower-end hotels most vulnerable to this increased competition. The scaling up of listings has also encouraged the professionalisation of short-term letting as an industry. It is now possible to outsource everything from house-cleaning to pricing to full-time service providers (Gopal and Perlberg 2015; Coldwell 2016), potentially decreasing the local impact of spending by hosts and guests. The more complex issue, however, is the impact of Airbnb on housing markets and residents. The primary concern is that landlords may find that short-term lets are more profitable than long-term lets to full-time tenants (Megginson 2015). In terms of housing market dynamics, Wachsmuth characterises Airbnb as opening up a 'rent gap' (Wachsmuth and Weisler 2017): the difference in yield between low-income, long-term tenants and higher-income, short-term visitors is driving eviction, renovation and gentrification.

Key drivers of Airbnb's housing market impact are always-available whole property listings, and hosts listing multiple properties. O'Neill and Ouyang (2016) found that multiple-unit operators and full-time operators make up a growing percentage of Airbnb hosts, and that between September 2014 and 2015 they generated US$500 million of Airbnb's total revenue of US$1.3 billion. In Los Angeles, Samaan (2015, p.2) found that nearly 90% of Airbnb's revenues were generated by hosts or leasing companies renting out either whole units or two or more units. Since high concentrations of Airbnb listings overlapped with areas with higher rents and lower rental vacancies, the report argued that renters are disadvantaged when these listings go to tourists. The report estimated that the 7,316 units were taken out of the rental market by Airbnb, equalling seven years' worth of affordable housing construction in Los Angeles (Samaan 2015, p.3). Similar finds have been reported by Gurran and Phibbs (2017) in the context of Sydney. In other cities it has been observed that 'superhosts' – those with three or more listings of entire homes – account for a disproportionate number of listings. In San Francisco, they account for 4.8% of all hosts, but operate up to 18% of the city's listings (Said 2015). In New York City up to 30% of Airbnb listings came from commercial hosts (BJH Advisors 2016).

How have cities responded to the issues raised by Airbnb?

Airbnb continues to operate without much oversight in many cities, but concerns that it is reducing the already constrained supply of rental housing have led some cities to increase regulation of short-term rental or ban Airbnb altogether. Approaches include limiting the number and duration of rentals, requiring Airbnb properties to be registered and imposing penalties for breaches (see O'Sullivan 2017). Community groups have also been formed in response to the perceived impact on housing affordability and residential amenity in popular Airbnb cities. Share Better Coalition (founded 2016) and Illegal Hotels Working Group (2015) in the US, and Neighbours Not Strangers (2016) in Sydney and We Live Here in Melbourne (Heaton 2016) are examples of groups of neighbours, community activists and/or elected officials working to highlight concerns about Airbnb, including rising rents, reduced public safety, neighbourhood disturbance and racial discrimination.

In response to these regulatory and community reactions, Airbnb has implemented a new policy of 'One Host, One Home' (Airbnb 2016b, p.7; Lehane 2016) in a handful of cities. The company also claims that they remove listings from 'unwelcome commercial operators' – generally hosts with multiple property listings – so that guests can have unique local experiences. The company has also launched 'home sharers' clubs, similar to local unions, to advocate on its behalf (Hickey and Cookney 2016; Dougherty and Isaac 2015). At the national and supra-national level, Airbnb is also part of a 47-company effort to convince European Union leaders that the sharing economy helps growth and should not be restricted by piecemeal national laws (Christie 2016).

Regulating Airbnb listings presents a twofold challenge. Firstly, cities face strong opposition from the company and the local advocacy groups it has organised. Second, even when a city introduces regulations for short-term rentals, enforcing the new rules may be impossible for authorities operating on a tight budget and with limited access to Airbnb data. There is a limited but growing body of research on how to overcome these regulatory challenges, including a number of Australian examinations (Minifie 2016; Lyons and Wearing 2015). At one end of the spectrum, Guttentag (2013) proposes fully legalising short-term rentals on sharing platforms, effectively integrating Airbnb into the formal economy to facilitate regulatory oversight and taxation. By contrast, Miller (2016) emphasises that while the sharing economy must be 'daylighted', as firms like Airbnb disrupt existing regulatory structures, they instead require novel regulatory responses. To that end, a model of transferable sharing rights is proposed, which would rely on algorithms and rental data shared by Airbnb and other platforms. Acknowledging that new technologies present new challenges, Miller (2016) suggests that technology should be harnessed to present new solutions. The barrier, however, is that many sharing firms resist public disclosure of data generated by their users.

What challenges do data issues present for cities?

In some cases, the data void has been filled by technically skilled members of the public: Inside Airbnb (Cox 2015) uses automated tools to 'scrape' property listings from Airbnb under a 'fair use' defence in order to provide advocates and policy-makers with data on Airbnb's impact: its data suggests that many more Airbnb listings are for entire homes than is commonly realised, and that many of these are rented permanently and, consequently, can disrupt housing supply and communities. This data has been used by several newspaper articles to detail the 'Airbnb effect' in cities such as San Francisco (Said 2015) and New York, as reported by Coldwell (2016) and others. Coldwell (2016) cites data indicating a rise in commercial listings on Airbnb and questions if it is really a 'sharing' economy if so many listings are commercial in nature.

Airbnb has questioned the validity of such data scrapes, suggesting that they draw negative conclusions about a small number of hosts (Said 2015). The consistent response from Airbnb is that the majority of hosts are middle-class families sharing their homes. Research commissioned by Airbnb indicates that, while it pushes up rents in major American cities, it does so only slightly (Kusisto 2015). Thomas Davidoff, the author of the report, suggests that criticism about Airbnb is about the right of people to stay in desirable places, rather than about housing affordability.

At the same time, however, the limited data produced by Airbnb to support these claims has been called into question. By comparing their scraped data with data released officially in an Airbnb's report, Cox and Slee (2016) found that Airbnb purged over 1,000 'entire home' listings for New York City immediately before the public release. This suggested an attempt to mask the true scale of whole-unit rentals in the city. Officials have responded with disapprobation: for example, New York State Senator Liz Krueger (Krueger 2016) claimed that this was evidence that Airbnb is unwilling to remove 'bad actors' from the site. Without definitive data, however, it remains challenging for cities to assess and address the impact Airbnb may be having on both its tourism accommodation and its housing markets.

Housing and tourism in Sydney, London and Phoenix

Before demonstrating how data access and analysis can help to tackle some of the regulatory challenges Airbnb poses for cities, it is helpful to provide a brief contextual overview of the three case study cities. This section outlines the key characteristics of Sydney, London and Phoenix, and provides a brief insight into the rise of and response to Airbnb in each. The three cities offer a useful combination of case studies, as each has reacted to the rise of Airbnb in different political and regulatory ways.

Sydney

Sydney is the capital of the state of New South Wales (NSW) and, with a population approaching five million, is the largest city in Australia. It also has Australia's most expensive housing market, and is a major tourist destination. Sydney's harbourside CBD, where most of the major cultural institutions are located, is a focus for tourists, with hotels concentrated in the CBD and inner suburbs. The peak tourist season is in summer, when Sydney's beaches (to the east and north of the CBD) are a major attraction. The CBD, inner suburbs and eastern suburbs are also popular for visitors on longer stays and working holidays, who often reside in backpacker hostels or shared flats, rather than traditional hotels. Comprising 35 local government areas, Sydney does not have a metropolitan government (the City of Sydney covers only the CBD and inner suburbs), leaving the state government to play this role.

Airbnb (2016a) reports that it supported A\$214 million of economic activity in Sydney in 2015 and that it was the suburbs that benefited most, including many areas that had not previously profited from tourism. As in other cities, the firm continues to emphasise that the 'overwhelming majority' of Sydney hosts are ordinary residents who rent out their homes. However, a *Sydney Morning Herald* analysis of Inside Airbnb data showed the number of listings has rapidly increased in recent years (Ting 2016). Moreover, the maps suggest Airbnb guests tend to stay in popular holiday locations like Bondi Beach, and not the less 'touristy' areas claimed by the firm. The regulation of short-term letting in Sydney is in a state of flux. In 2016 a NSW Parliamentary Committee conducted an inquiry into the adequacy of the regulation of short-term holiday letting (see NSW Parliament 2016). In relation to current planning instruments, use of a private dwelling for short-term letting may or may not involve a change in its use. There is also no clear threshold for determining a change of use, nor a consistent definition of short-term letting across Sydney's local government councils.

There is also uncertainty about the private regulation of short-term letting, particularly in apartment buildings in strata title schemes. Under strata title legislation, owners corporations can make wide-ranging by-laws about activities in apartments, but cannot prevent apartment owners from leasing or otherwise 'dealing' with units. This casts doubt on the validity of by-laws prohibiting or restricting short-term letting (see, for example, *Estens v Owners Corporation SP 11825* [2017] NSWCATCD 52). At the same time, however, the Committee found few complaints regarding short-term letting, with the complaints heard relating principally to residential amenity. On the issue of housing affordability, the Committee found no robust measurement and data collection to indicate that short-term rentals are removing properties which would otherwise house long-term tenants (NSW Parliament 2016, p.12) That said, the report fails to examine the issue at a localised level, instead focusing on the percentage of Airbnb properties Sydney-wide.

Airbnb emphasises that they are willing to work with new legislation in Sydney or elsewhere. Their submission to the Parliamentary Committee notes that 52% of their hosts live in low to moderate income households, while Airbnb guests are often highly educated professionals (Orgill 2015, p.5). This is consistent with Airbnb's framing of hosts internationally, which normalises individual hosts while ignoring the prevalence of commercial landlords on the site. Airbnb recommends that the NSW Government put in place a state-wide policy to allow individuals who occasionally rent out their home to do so without approvals or licences. The submission also recommends that councils have the flexibility to decide on commercial holiday homes.

Overall, the Committee's reticence to take a tougher stance on Airbnb appears to reflect the NSW Government's support for the 'collaborative economy'. In a position paper (NSW Finance Services and Innovation 2016), the Government emphasises the opportunities that the sharing economy presents for consumer choice, employment and productivity. Since the Committee's report was handed down, however, the NSW Department of Planning has produced an Options Paper on short-term letting, and invited public comments. While it has yet to report back, a department representative noted that the submissions received number well into the thousands.

London

London is the UK's capital and largest city, and an international financial and cultural centre. One of the world's most popular tourist destinations, it has a very expensive housing market. London is governed by the Greater London Authority under the Mayor, as well as by 33 local authorities. For decades London has specifically regulated short-term letting, with the objective of preserving housing stock; recently it has loosened its regulations. Until 2015, the Greater London Council (General Powers) Act of 1973 restricted short-term letting of properties within Greater London with a view to preventing the spread of 'temporary sleeping accommodation'. The act applied to both entire home and partial home lets, and determined that a lease for less than 90 consecutive nights was a short-term let requiring a formal change of planning use. Failure to properly follow this process would prompt a planning enforcement notice to be served to the person carrying out the lease (Simcock and Smith 2016). In practice, however, the rise of home-sharing/letting platforms and weak enforcement meant that many homeowners operated in ignorance, or defiance, of the law.

The Deregulation Act 2015, Section 44, enacted in 2015, relaxed these restrictions, which had become controversial following the rise of Airbnb and, in particular, the tourist influx around the London 2012 Olympics. Under the new law, almost any property could be leased on a short-term basis provided that the annual aggregate did not exceed 90 nights (www.legislation.gov.uk/ukpga/2015/20/section/44/enacted). In principle, the regulation applies to both whole home and room rentals, but in practice enforcement by both Airbnb and councils has focused solely on whole-house lets: Airbnb automatically caps entire home listings at 90 days *unless* the host presents appropriate documentary evidence of permission to let for longer. The new regulations seek to balance flexibility with respect to the 'sharing economy' against a shortage of properties in the long-term rental market. Short-term letting is also privately regulated, with most of London's apartments being 'owned' under long-term leases. These leases often include terms prohibiting the use of apartments other than as a 'private residence', and in a number of cases short-term letting has been found to be in breach of these terms, resulting in termination of the lease (Nemcova v Fairfield Rents Ltd [2016] UKUT 303 (LC) (6 September 2016)).

On the whole, the UK remains supportive of both Airbnb and the wider sharing economy; Business Minister Matt Hancock is quoted by Airbnb as saying:

> Platforms like Airbnb are empowering a generation of innovators and everyday entrepreneurs. They are disrupting the status quo and making sure consumers get the very best deal. We will back them all the way and continue to remove barriers to their success.
> *(www.airbnbcitizen.com/queen-signs-home-sharing-into-uk-law/).*

Within London, homeowners (and lessees) have responded in a variety of ways. While many comply with the law, others are listing rooms individually for 365 days of the year, even if every room in the property is available to let, and some are (illegally) listing the same property across multiple platforms such that no one channel exceeds the 90-day limit (Lynn and Allen 2017). At a time of raging controversy over overseas buyers and empty flats, there is growing concern that not only are the regulations being flagrantly ignored by owners who have little to fear from council enforcement, but that the 'Airbnb effect' is driving the rapid gentrification of historically deprived Inner London neighbourhoods such as Hackney.

Phoenix

With a resident population estimated at four million in 2015, the Phoenix metropolitan area (PMA) is the fifth largest in the US (US Census Bureau n.d.). The PMA comprises 24 municipalities including the cities of Phoenix, Scottsdale, Mesa, Glendale and the retirement communities of Sun City and Sun City West. Phoenix was one of the hardest hit areas during the last housing bubble, although 2017 has reportedly been one of the better years for home sales since 2008 (Reagor 2017). This increase in home sales has been attributed to rising rent prices which are driving residents to move to homeownership. The Home Buying Institute has predicted that available homes will fall short of demand from 2018, causing a higher level of competition among buyers (Cornett 2017).

Situated in the Sonoran Desert of the Southwest US, the area experiences low rainfall and mild temperature winters. Amenities on offer include luxury golf courses, high-end shopping, resort spas, viewing spring training for Major League Baseball and a number of culturally significant museums. These benefits combine to attract both seasonal visitors – the region hosts a large 'snowbird' population, who seasonally migrate to the area to enjoy the mild winters – and other short-term visitors. Scottsdale is reported to host the most out-of-town visitors, but is closely followed by Phoenix, which is often used as a base of operations for tours to Sedona and the Grand Canyon. In 2016, 43 million people visited the state of Arizona as a whole, spending over US$21 billion, which resulted in jobs and significant tax revenues for the state residents (Dean Runyan Associates 2017).

Phoenix adopted a *laissez-faire* stance to Airbnb, requiring only that the company arrange automatic tax remittance on behalf of hosts (Lines 2015). With broad interest in supporting a shared economy, and to maintain support of the tourist industry, the State of Arizona ratified Senate Bill (SB) 1350 in 2016, which effectively stops local municipalities and cities from banning short-term rentals (Etienne 2016). The impact is 7,600 new Airbnb hosts in Arizona – a 61% net growth from 2015 to 2016 (*Phoenix Business Journal*, Totten 2017). Those opposed to SB 1350 felt that municipal bans were needed to (a) protect homeowners concerned about the impact on home values and (b) minimise the financial impact on hotels and resorts.

While Arizona municipalities are unable to ban short-term home rentals, SB 1350 does not prevent Homeowner Associations (HOA) from implementing a neighbourhood ban. HOAs are neighbourhoods self-governed through Covenants, Conditions and Restrictions (CCRs) – not altogether unlike the rules governing Sydney's strata – that aim to protect property values and include requirements such as paint colours, landscaping guidelines, parking and pet ownership. In the US Southwest (including both Arizona and southern California), there are approximately five million housing units in HOAs. While the extent of Airbnb restrictions in HOAs is currently unknown, there is a potential to see increasing restrictions across many parts of the PMA.

Approach to spatial analysis

As noted above, the aim of this chapter is to explore how spatial data analysis tools can help to support cities in responding to the rise of Airbnb in an equitable and efficient way. Given that many of the debates of Airbnb's impact relate to where listings are located, one way to achieve this is to map the spatial patterns of Airbnb properties. In particular, this involves quantifying and visualising the proximity of listings to city centres, and examining hot spots in detail. Ultimately, we want to determine whether the distribution of Airbnb properties provides evidence of neighbourhood impacts and of a shift from long-term to short-term lets in inner city areas. But the insertion of space into a quantitative analysis complicates things substantially because of issues of scale; for instance, globally Airbnb listings may cluster within the CBD, but within the CBD they might be fairly evenly distributed. In contrast, although there may be far fewer listings in suburban areas, they may be heavily clustered around public transit. In short, clustering at one scale does *not* imply clustering at other scales.

In addition, we also have to choose between treating each listing as a point in space, or grouping the listings together based on whether they fall within the area that defines an administrative or statistical region (e.g. Census tract, postcode or borough). Treating each listing as a point enables us to examine variation at a finer scale and higher resolution, but it can also make it harder to spot larger patterns since local variation can mask overall trends. Again, there is no 'correct' answer as to which mode of analysis is best: the choice depends very much on the purpose of the analysis and the suitability of the data. For a neighbourhood impact analysis it makes sense to try to link each listing to a local statistical area: not only does it enable us to derive some measure of an *average* listing, it also allows us to bring in government data on the area's demographics.

Methodology

The analysis was performed on Airbnb listings recorded between 23 August and 22 September 2016 for Sydney, London and Phoenix. The data provided by AirDNA allowed the researchers to differentiate Airbnb listings into two types of properties and two types of hosts. Traditional Holiday Lets are defined as entire property listings that are leased for more than 90 days. House Share listings are entire property listings that are leased for no more than 90 days, or private and shared room listings of any duration. If several listings shared the same host ID they were counted as Multi-listings regardless of whether they were for separate properties or separate bedrooms/bedspaces within the same property. All other hosts were classed as Non Multi-listings, though it should be noted that some Multi-listings could also represent professional hosting organisations that list and operate properties on behalf of individual owners. For the purposes of this analysis we take all multiple listings to be indicators of commercially orientated rentals, even if the property itself is not owned by a commercial entity.

The four analytical techniques used are as follows:

Dot density

Dot density maps have been created for each of the three cities for the Airbnb listings, each representing a distinct point feature, using ESRI's ArcGIS software. The dot density maps illustrate those areas where there is concentrated or sparse Airbnb activity. The dot density maps comprise symbology which differentiates traditional holiday lets and house shares; all Airbnb listings are either/or, so the combination of these two types of listings should give a visual impression of the distribution of listings throughout the three metropolitan areas.

Buffer analysis

Buffer analyses were conducted around the CBDs of Sydney and London, and Phoenix's downtown area, and we summarise the proportion of Traditional Holiday Lets, House Shares and Multi-listings falling within each buffer zone. A range of buffers – with spacings of 500m, 1km, 2km, 5km and 10km – were generated from the boundary of the CBD/downtown area using ArcGIS's multiple buffer rings tool.

Location quotient analysis

The location quotient (LQ) is a simple, normalised measure of concentration, first used in the 1940s (see Florence 1948 and modifications by Hall 1962). Although some analytical issues have been noted with the original approach (e.g. O'Donoghue and Gleave 2004), it is still frequently employed in economic geography and regional analysis. In particular, the LQ has the benefits of setting a clear baseline expectation for the distribution of factories or listings, and of being easy to interpret in that a value of one indicates that the proportion of employment (or Airbnb multi-host listings) in Zone A is the same as that of the proportion of employment (or Airbnb multi-host listings) in the overall Region R. LQs less than one indicate lower densities than in the general region, and greater than one indicates higher densities.

In this research a primary interest is to understand how the balance of whole-unit to shared-unit listings varies by area and to benchmark this against the region as a whole, because different cities have different overall proportions to begin with. By normalising the data using the LQ we make the results more readily comparable.

Moran's I

Anselin's Local Moran's I is defined as 'a cluster and outlier analysis that can be run on a set of weighted geographical features (our census geographies). The process generates an index *I* for each feature' (Anselin 1995, p.94).

A positive I indicates that a feature has neighbours that have similar values to it, thereby detecting clusters. A negative I indicates that a feature has neighbours with dissimilar values, thereby detecting outliers. Features with high values (LQs for either Traditional Holiday Lets or Multi-listings) that are surrounded by similar high values are High-high clusters or hot spots. Features that have neighbours with similar low values are Low-low clusters or cold spots. Features that have high value and have neighbours with low values are considered High-low outliers, and the inverse are considered as Low-high outliers (Mitchell 2005).

For this analysis, hot spot maps for both Traditional Holiday lets and Multi-listings have been generated using the Cluster and Outlier Analysis tool in the ArcGIS Spatial Statistics Toolbox. 'Neighbours' were defined as polygons that share edges and corners.

Limitation

Census-derived geographies were used as proxies to define 'city centres' and 'downtown areas' for the buffer analysis, and to aggregate the raw AirDNA point data into 'localities' for the tandem LQ-Moran's I hot spot analysis. The SA2 geography was used for Sydney; MSOA or Middle Layer Super Output Area for London; and Census Tracts for Phoenix.

Employing a uniform way to define the boundaries of localities and the 'city centre' for Sydney, London and Phoenix is complicated given that the three metropolitan areas vary widely in size, structure and density of activity. In some cases, what has been identified as the city centre may not necessarily represent the focal point of tourist activity in reality, and what can be popularly regarded as a locality may not be fully captured by the boundaries of census geographies. This may be an exercise in itself for future research work to improve on. Ultimately, the decision to use Census geographies for this chapter was motivated by their availability and that they also allow the appendage of a range of household data that may be useful for future research work.

Findings

London, Sydney and Phoenix vary enormously in size and structure, with Sydney and London having more defined central business areas compared with Phoenix, which is far more dispersed. Similarly, the areas vary: 100% of Airbnb listings for London can be found within 30km of the CBD, while for Sydney that distance is 70km; for Phoenix, while 98% of listings can be found within 50km of the downtown, some are as far away as 160km. There are also significant differences in terms of the type and number of properties listed in each city; although shared rooms are a small proportion in all cities, the balance between whole units and private rooms varies substantially with the regulatory, financial and morphological context. Table 3.1 summarises Airbnb listings by type for each city, showing that entire properties are much more common in Phoenix than in London (though with a much lower count), while Sydney and London are broadly comparable in terms of total properties but differ in terms of the balance between whole and partial lets.

The balance between Traditional Holiday Lets and House Shares (Table 3.2), and between Multi-listings and Single property hosts (Table 3.3) helps to demonstrate why comparative spatial analysis can be useful: between the three case studies there are three different mixes of full- and part-time lets, and (to a lesser extent) between professional and non-professional hosts; however, it's impossible to tell from these overall distributions whether these differences are meaningful, and whether they have any bearing on our understanding of the 'Airbnb effect'.

TABLE 3.1 Airbnb listings by type for each city

Listing type	Greater Sydney		Greater London		Metropolitan Phoenix	
	Listings count	% of total	Listings count	% of total	Listings count	% of total
Entire property	13,239	60%	11,626	52%	4,289	71%
Private room	8,293	38%	10,452	46%	1,711	28%
Shared room	382	2%	411	2%	71	1%
Total	21,914	100%	22,489	100%	6,071	100%

TABLE 3.2 Summary of Traditional Holiday Lets vs House Share listings

	Greater Sydney		Greater London		Metropolitan Phoenix	
	Listings count	% of total	Listings count	% of total	Listings count	% of total
Traditional Holiday Let	5,083	23%	3,832	17%	2,642	44%
House Share	16,831	77%	18,657	83%	3,429	56%
Total	21,914	100%	22,489	100%	6,071	100%

TABLE 3.3 Summary of Multi-listings vs Single listings

Multiple listings	Greater Sydney		Greater London		Metropolitan Phoenix	
	Listings count	% of total	Listings count	% of total	Listings count	% of total
Yes	6,924	32%	9,190	41%	2,422	40%
No	14,990	68%	13,299	59%	3,649	60%
Total	21,914	100%	22,489	100%	6,071	100%

Dot density maps

The three maps below provide both an overview of the distribution of properties by type across each metropolitan area: (a) each dot in these maps represents an Airbnb listing; (b) the dots represent listings that have been categorised as Traditional Holiday Let; (c) the dots represent Multi-listings, which are listings that have been identified to share common hosts. To help orientate readers who might be unfamiliar with these cities, the Sydney and London CBDs have been outlined in black. The same has been done for the downtown areas of Phoenix and Scottsdale for the Phoenix Metropolitan Area (PMA).

Buffer analysis

The tables and graphs below show the results of building a buffer around the CBD using increments of different sizes (as shown in Figures 3.1, 3.2, 3.3): 500m, 1km, 2km and 5km. Although a fixed size would give a more readily interpretable 'standard' distance from the downtown, the varying width allows us to unpick the 'city fringe' (border of the CBD) from older 'inner city' suburbs, and the higher-density commuter suburbs from the lower-density outer metropolitan area. This approach allows us to look for systematic variation in the types of hosts and properties using a more meaningful 'region'.

Sydney

In the context of Sydney (Figure 3.4), the share of traditional holiday lets versus that of house share listings shows a decreasing trend as one moves away from the CBD area. And while the raw count of traditional holiday lets is greatest in the outer distance bands – probably as a

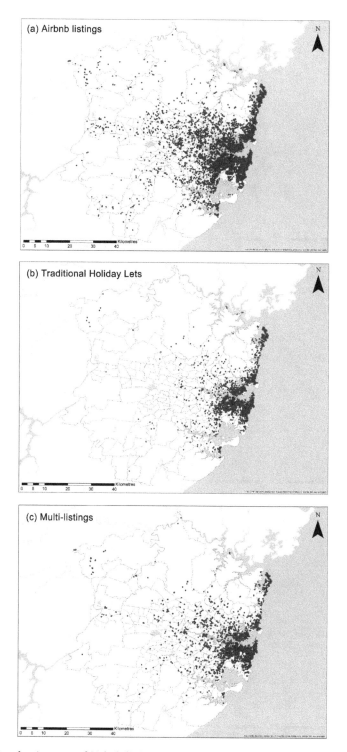

FIGURE 3.1 Dot density map of Airbnb listings across Greater Sydney

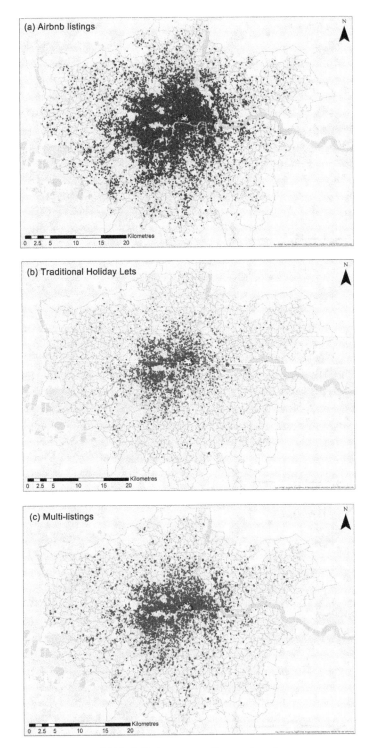

FIGURE 3.2 Dot density map of Airbnb listings across Greater London

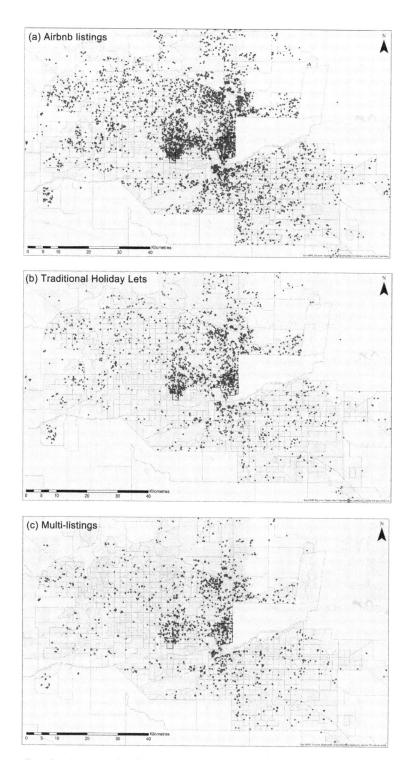

FIGURE 3.3 Dot density map of Airbnb listings across Phoenix Metropolitan

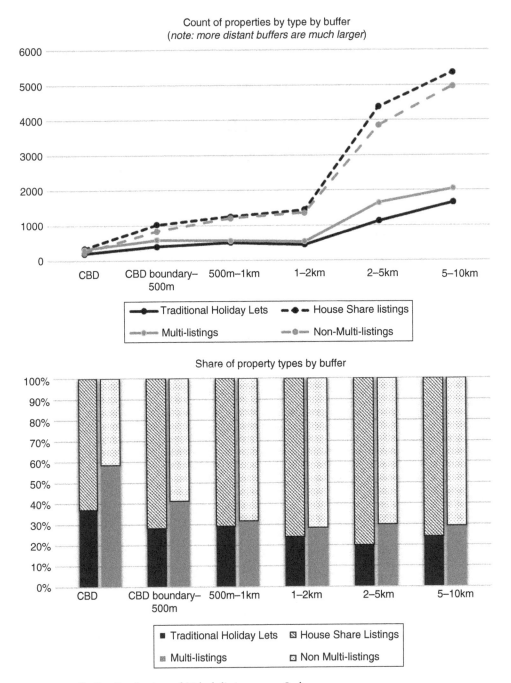

FIGURE 3.4 Buffer distribution of Airbnb listing across Sydney

result of the outer buffers covering much more area – the greatest proportion of traditional holiday lets is found within the CBD where 37% of the listings fall within the category. The same decreasing trend in share can be observed for multi-listings; where the majority (58%) of Airbnb listings inside the CBD are categorised as multi-listings, the majority of listings in each of the outer buffer zones are categorised as non multi-listings.

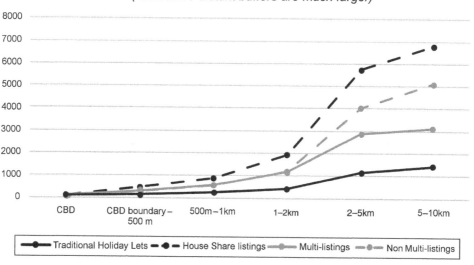

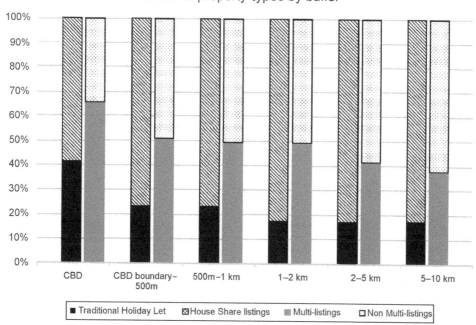

FIGURE 3.5 Buffer distribution of Airbnb listing across London

London

Within the CBD, 42% of the listings are categorised as traditional holiday lets. This represents the greatest proportion of traditional holiday lets versus house share listings across six buffer zones (Figure 3.5). Similar to Sydney, the shares of both traditional holiday lets and multi-listings demonstrate a decreasing trend as one moves away from the CBD. However the proportion of share of multi-listings in London are greater than what is seen in Sydney. Multi-listings

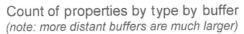

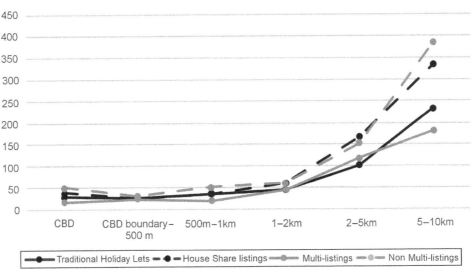

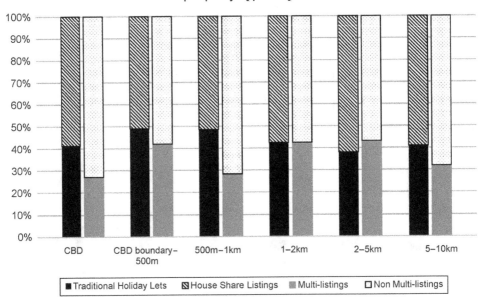

FIGURE 3.6 Buffer distribution of Airbnb listing across Phoenix

account for half of all listings occurring within 2km outside the CBD area, dropping down to 42% and 38% in the two outermost buffer zones, respectively.

Phoenix

The data for Phoenix tells a somewhat different story, as illustrated in Figure 3.6. The shares of both traditional holiday lets and multi-listings are in the minority compared to house share listings and non multi-listings, respectively. The shares of traditional holiday lets fluctuate from 38–49% over the six buffer zones, with the biggest proportions (49%) occurring within the first 500m outside downtown Phoenix and from 500m to 1km. In the cases of Sydney and London, a radial pattern could be discerned by buffer analysis centred around the CBD. This buffer approach may be limited in the case of Phoenix where there are different downtown areas that could serve as multiple centres of Airbnb activity. In the dot density map for Phoenix, traditional holiday lets are apparently denser in the areas around downtown Phoenix, Oldtown Scottsdale, and a little way north from Oldtown Scottsdale.

Hot spot and location quotient analysis

In this analysis the focus in on traditional holiday lets and multiple listings, as these are considered to be two proxies of where Airbnb listings could be impacting traditional housing supply in taking them off the long-term rental market. Location quotient analysis was used to evaluate the incidence of traditional holiday lets and multi-listings in a way that is comparable to the extent of Airbnb penetration within Sydney, London and Phoenix. The location quotients were then analysed using Anselin's Local Morans I to locate clusters of hot and cold spots for traditional holiday lets and occurrence of multi-listings (Figures 3.7–3.12).

Sydney

Through the Moran's I analysis, it is observed there are two key High-high clusters of Airbnb activity for traditional holiday lets. The first includes the CBD and North Sydney and extends along the harbour to the coast at Bondi. The second cluster is focused around Palm Beach. These high cluster areas are associated with the main tourist destinations of Sydney, located either in or adjacent to the CBD or harbour and beach fronts. These areas also represent some of the most unaffordable housing market locations for Sydney and are also areas with comparatively higher rates of rental housing.

There were cold spots identified in the outer west; however, these areas include significant tracks of undeveloped land, some of which has been identified for future urban growth. The composition of the spatial boundaries used in these fringe areas likely accentuates the 'cold spot'. Despite this problem, these areas being relatively far from the economic centre of Sydney would also account for lower rates of Airbnb listings. LQ analysis confirmed these patterns, although it identified an extended area of influence of Airbnb listings to cover the broader northern Sydney area including Manly, and extended south along the coast to include Bondi, another popular Sydney beach and tourist destination.

There is no significant clustering of multi-listings in the areas highlighted above. However, the area of Epping shows a High-high clustering and the LQ shows a corridor of listings of this type spanning from the CBD out to Epping. The reasons for clustering in Epping may relate to its proximity to the Macquarie business park and Macquarie University, and its strategic location on Sydney's railway network. Both this cluster and LQ results require further qualitative analysis to understand the dynamics of these outputs.

London

Overall, London shows high concentrations of Airbnb activity within the CBD and the adjacent boroughs of Westminster, Kensington and Chelsea, Camden, Islington, as well as across

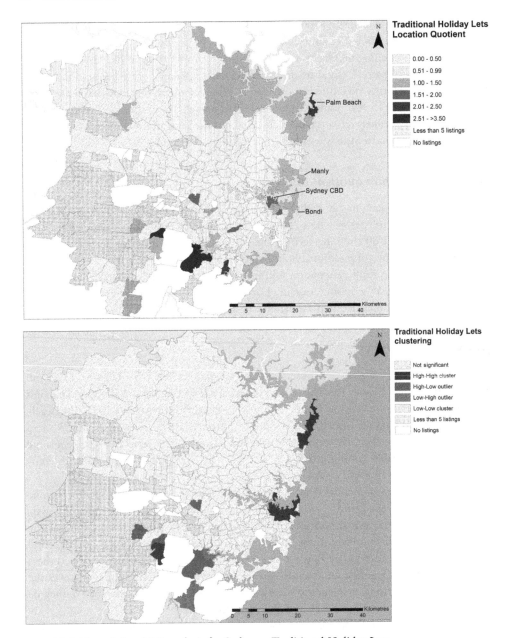

FIGURE 3.7 Moran's I and LQ analysis for Sydney – Traditional Holiday Lets

the river to the northern parts of Southwark. This is very close to a standard definition of 'Inner and Central London', a mix of extremely wealthy and disadvantaged areas in which gentrification has long been an issue.

In contrast, the less accessible parts of Outer London are – not altogether surprisingly – cold spots for Airbnb. Some of these areas could be more than an hour by public transit from the main tourist attractions, and it is largely for this reason that hot spots for traditional holiday lets in London seem to be clustered around Inner and Inner-West London. Following

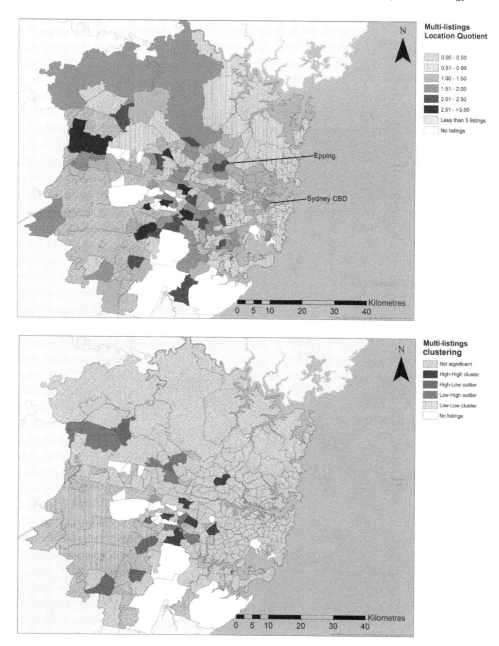

FIGURE 3.8 Moran's I and LQ analysis for Sydney – Multi-listings

the Thames upstream and to the south-west, we also find some evidence of 'holiday lets' in Hammersmith and Fulham, Richmond and Kingston upon Thames. However, in conjunction with the 'hot spot' near City Airport to the east of Canary Wharf, this should give some pause for thought about what, exactly, is being shown. Although it would require further research, one explanation is this area may provide temporary accommodation to highly mobile business elite catering nearby financial services.

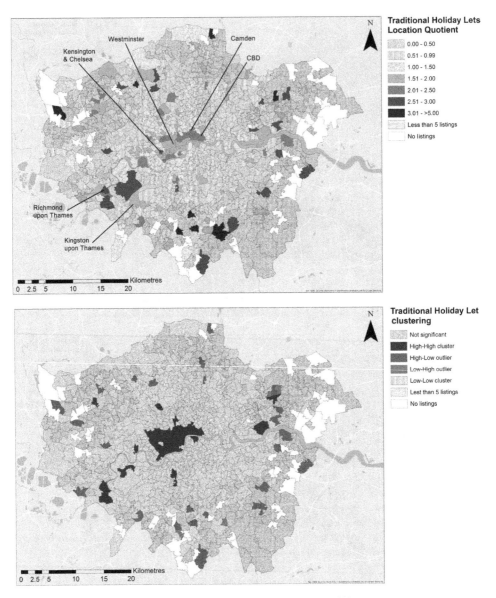

FIGURE 3.9 Moran's I and LQ analysis for London – Traditional Holiday Lets

The hot spots shown for multi-listing hosts reinforce the interpretation of the earlier maps, while also deepening our appreciation of contrasting trends outside the core tourist areas. In 'prime' West London (including the West End and parts to the south-west) there is a strong incentive to professionalise the 'sharing economy' – not only are many properties let as a whole (meaning that they are empty when not rented), but they are also owned or managed by 'hosts' who operate multiple properties. The same dynamic does not appear to apply to the same extent to the north or south-west of the CBD in historically residential areas such as Islington or Richmond, which seem to contain both fewer 'holiday lets' and fewer multi-listings.

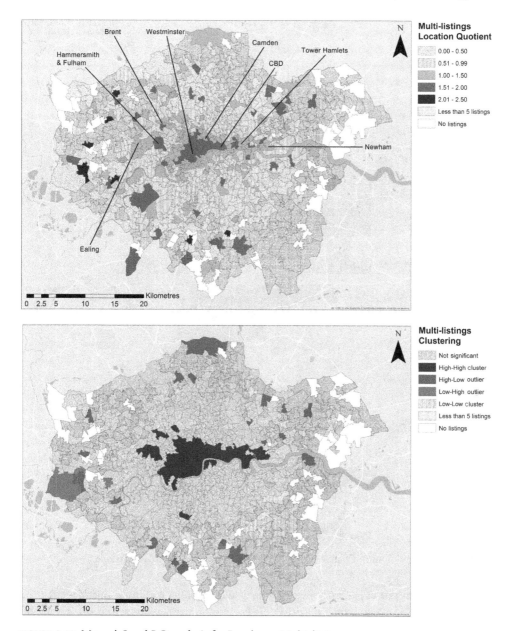

FIGURE 3.10 Moran's I and LQ analysis for London – Multi-listings

What is rather more unexpected, however, is the extension of the central cluster of hot spots towards Tower Hamlets and Newham. We might expect there to be some impact from the business district of Canary Wharf – particularly if 'holiday lets' includes business-travel catering – but this seems much wider in extent at first glance. Although the geographies do not fully overlap, suggesting slightly different dynamics in operation, there is clearly an alignment of increased listings, increased professionalisation and increased market stresses in trendy Inner East London.

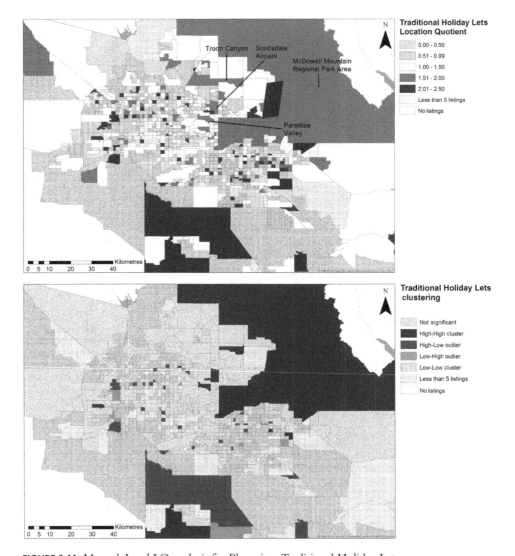

FIGURE 3.11 Moran's I and LQ analysis for Phoenix – Traditional Holiday Lets

However, from this exploratory overview it would be difficult to tease out the extent to which the increased rate of listings is a driver of gentrification or a response to it. Obviously, the 'rent gap' between long- and short-term rentals in Hackney and its environs increases the likelihood that owners will withdraw from the long-term rental market, and consequently drive up the price of the remaining listings. However, it is also highly likely that no small number of listings are being driven by landlords seeking to cover mortgage payments on properties in which they live, as well as tenants (often) illegally letting out a room in order to 'make rent'.

Phoenix

The hot spot maps show clustering of traditional holiday letting in areas surrounding Scottsdale airport, the affluent neighbourhood of Paradise Valley and Troon Canyon. To an extent, this

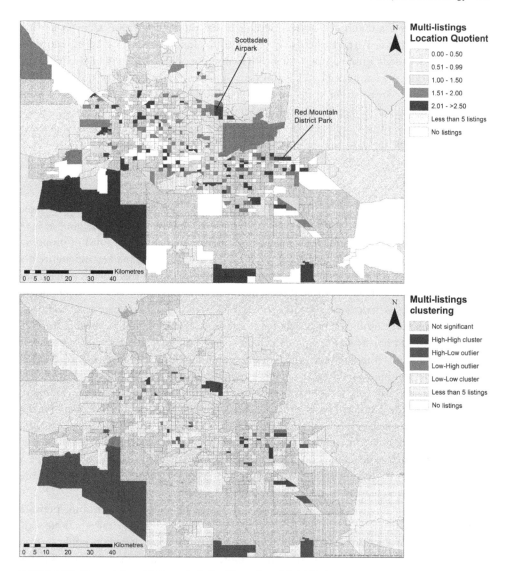

FIGURE 3.12 Moran's I and LQ analysis for Phoenix – Multi-listings

result is congruent with the earlier dot density map where there is an apparent concentration of listings to the north of Oldtown Scottsdale. Traditional holiday letting may be expected to cluster around the CBD in Sydney and London, because this is where many tourist spots and major public transport interchanges are co-located. In Phoenix, however, the same expectation may not be so readily applied. While downtown areas also attract concentrations of Airbnb listings, other tourist spots such as golf courses and ball parks are dispersed throughout neighbourhoods of Phoenix and Scottsdale. Proximity to public transport hubs may also be less of a factor for Airbnb guests and hosts in metropolitan Phoenix, where vehicular transport may be more of the norm and where road and traffic conditions are apparently ideal for piloting a driverless taxi service (Lee 2017).

The hot spots for traditional holiday letting in Paradise Valley, Troon Canyon and the surrounds of Scottsdale Airpark, then, may be due to the luxury offered by Airbnb listings in these neighbourhoods. Entire property lets with swimming pools, patios and proximity to golf parks characterise the results of a quick Airbnb search in these areas. It must be noted, however, that the downtown areas cannot be discounted as potential hot spots. The method used for this hot spot analysis involved aggregating Airbnb listings into Census tracts. Using an alternative aggregation method that can better reflect a coherent neighbourhood may be used in future analyses to improve the detection of hot spots.

Analysis

Overall for Sydney, the data indicates there are around 22,000 Airbnb listings and nearly one third of these, almost 6,000, can be defined as traditional holiday lettings, and almost 7,000 of the total properties are multiple listings. These findings suggest over a third of properties listed on Airbnb are potentially been taking off the long-term rental market and are used as short-term lets. In an already unaffordable housing market the loss of this stock is problematic. Unsurprisingly, this activity as expected is occurring in and round the CBD and spanning out to tourism hot spots, specifically Bondi, Coogee and Manly Beaches. There is an additional hot spot in the north around Palm Beach. There is also a corridor of multiple-listings spanning from the CBD out north-west to Epping, which includes the Macquarie Technology Park.

One of the most-visited cities in the world, London highlights the impact of the 'Airbnb effect' on cities: within Central London (very roughly, <2km from CBD) there are thousands of 'Traditional Holiday Lets' for which the returns are clearly sufficient to encourage property owners to incur the additional costs of letting their entire flats or houses year-round. Further out, in more residential areas, there is little in the way of clustering (fewer hot spots) except in the vicinity of major tourism and business attractions, suggesting that the impact of London's regulatory approach to dealing with landlords is substantial. The legal constraints on change of use (for more than 90 days a year) and the additional oversight associated with full-time letting do not seem to be as worthwhile outside the core.

However, the prevalence of multiple listings in areas where London's gentrification pressures are thought to be strongest *is* a matter of some concern. Multiple listings clearly indicate some form of professionalisation outside the traditional tourism market: it is impossible to tease apart whether these are landlords putting one or more rooms in one or more properties online so as to increase yield, or companies providing a kind of management service for hosts in which they deal with listing (including high-quality photography), cleaning and other administrative tasks in exchange for a slice of the rents. Either way, these point towards spaces that *might* otherwise be listed on the long-term rental market and, consequently, their presence in markets such as Hackney and Lambeth and Southwark requires further investigation.

For Phoenix, the storyline is clear. The data indicates over 6,000 Airbnb listings with close to half of them as traditional holiday letting. The listings are distributed across a large portion of the metropolitan area. While Phoenix is the anchor city for the metro area, the other PMA municipalities, such as Mesa, Scottsdale and the growing communities of Goodyear and Glendale, create a polycentric metropolitan area that reduces the strength of a Phoenix-only CBD. The polycentric structure of the PMA impacts the results of a spatial analysis when studying economic structures of tourism, because the location of the golf courses, resort spas and baseball fields are spatially dispersed throughout the region. The baseball fields alone – home to Major League Spring training – can be long distances from the Phoenix CBD. For example, the Goodyear Ball Park, which is the spring training location for the Cleveland Indians, is 35km from downtown Phoenix.

Nevertheless, the tourism story is poorly captured in this analysis because the data for this study were collected from the end of August until the end of September. During this time, schools are back in session – reducing the number of families who use Phoenix as base of operations for visiting the Grand Canyon and other northern Arizona destinations; the average daytime (and night-time) temperatures in August and September remain around 39°C – unappealing to golfers and outdoor enthusiasts; and baseball spring training occurs in March – eliminating the impact of sporting events to attract out-of-town visitors. So while the data shows the wide spatial variability of Airbnb in Phoenix, the true impact of this disruptive technology requires additional data and analysis.

Conclusions

In an era of rapidly emerging big data, smart cities and disruptive technologies, achieving equitable and efficient regulatory outcomes requires that city governments have access to an 'equal playing field' in terms of data access and analysis. As this chapter has demonstrated, there are various spatial analysis approaches that can be adapted to examine the impact of new disruptive technologies on urban space. In this paper we used Airbnb listing data provided through AirDNA to undertake spatial proximity analysis using dot density maps, buffering and hot spot analysis using location quotients and Moran's I. The outcomes show a spatial trend of more Airbnb listings closer to the central city for Sydney and London, and an apparent concentration of listings around the airport and affluent neighbourhoods in the case of Phoenix.

These approaches offer some useful insights, but also come with clear limitations. Interpreting the findings emerging from these forms of spatial analysis requires an understanding of how these tools are designed, as well as the particular characteristics of the cities examined. In this respect, the analysis outlined here highlights why spatial data analysis is best used to enhance, rather than replace, on-the-ground knowledge of cities and their residents.

To expand on the preliminary findings outlined here, further spatial analysis of Airbnb should focus on a finer scale, as well as considering proximity to major tourist attractors. Work by Gurran and Phibbs (2017) demonstrates that in Sydney, for example, there is a need to closely unpack the nature of significant Airbnb activity happening in tourism destinations like Manly and Bondi. Further research should also look at spatially overlaying housing affordability indicators and other spatial proximity factors, such as distance-to-transport interchanges and areas of high employment. Layered analysis of this kind will help to highlight the neighbourhood-level complexities of housing and tourism markets in these cities, which are far from uniform. In this way, these tools can provide valuable insights for governments seeking to respond to disruptive technologies like Airbnb as equitably and efficiently as possible.

References

Airbnb (2015). Queen signs home sharing into UK Law. Retrieved from: www.airbnbcitizen.com.
Airbnb (2016a). New study: Airbnb community contributes AUD $214 million to Sydney and its suburbs, brings tourists to new neighbourhoods. Retrieved from: www.airbnb.com.au/press/news/new-study-airbnb-community-contributes-aud-214-million-to-sydney-and-its-suburbs-brings-tourists-to-new-neighbourhoods.

Airbnb (2016b). One Host, One Home: New York City. Retrieved from: https://1zxiw0vqx0oryvpz3ikczauf-wpengine.netdna-ssl.com/wp-content/uploads/2016/07/OneHostOneHomeNewYorkCity-1.pdf.

Airbnb (2017): Generating $6.5 billion for restaurants around the world. Retrieved from: www.airbnbcitizen.com/2017restaurantreport/.

Alexander, J. (2015). Airbnb, social media and the quest for the authentic urban experience. The Conversation.

Anselin, L. (1995). Local Indicators of Spatial Association-LISA. *Geographical Analysis*, 27, 93–115; doi.org/10.1111/j.1538–4632.1995.tb00338.x.

Australian Associated Press (2016). Airbnb and Stayz call for consistent regulation for short-term holiday rentals. *The Guardian*.

BJH Advisors (2016). Short changing New York City: The impact of Airbnb on New York City's housing market. Housing Conservation Coordinators.

Brown, M. (2014). Airbnb: The growth story you didn't know. GrowthHackers.

Castells, M. (2010). *The Rise of the Network Society*. Chichester, West Sussex, UK: Wiley-Blackwell, 1–25.

Chen, Y. and Xie, K. (n.d.) Consumer valuation of Airbnb listings: a hedonic pricing approach. *International Journal of Contemporary Hospitality Management*, 29, 2405–2424; doi.org/https://doi.org/10.1108/IJCHM-10-2016-0606.

Christie, R. (2016). Airbnb leads call for EU to block nations' sharing-economy laws. Bloomberg.

Coldwell, W. (2016). Airbnb: from homesharing cool to commercial giant. *The Guardian*.

Copley, T. (2016). Regulating the sharing economy. Inside Housing.

Cornett, B. (2017). Phoenix Housing Forecast Through Summer 2018: Getting Back to Normal? Home Buying Institute.

Cox, M. (2015). How is Airbnb really being used in and affecting the neighbourhoods of your city? Inside Airbnb.

Cox, M. and Slee, T. (2016). How Airbnb's data hid the facts in New York City. Inside Airbnb.

Danneels, E. (2004). Disruptive technology reconsidered: a critique and research agenda. *The Journal of Product Innovation Management*, 21, 246–258; doi.org/10.1111/j.0737-6782.2004.00076.x.

Dean Runyan Associates (2017). Arizona travel impacts by legislative district (No. 2016p). Arizona Office of Tourism.

Deloitte (2015). Review of the collaborative economy in NSW. NSW Department of Finance, Services and Innovation.

Dougherty, C. and Isaac, M. (2015). Airbnb and Uber mobilize vast Uber base to sway policy. *The New York Times*.

Duke, J. (2016). Sydney's slumlord solution: big data to crack down on illegal boarding houses. Domain.

Duke, J. and Nicholls, S. (2015). NSW government investigating Airbnb rentals. Domain.

Dwoskin, E. (2016). Airbnb is forming an alliance with one of the nation's biggest labor unions. *The Washington Post*.

Edelman, B., Luca, M. and Svirsky, D. (2016). Racial discrimination in the sharing economy: evidence from a field experiment. Harvard Business School NOM Unit Working Paper No. 16–069; doi.org/10.2139/ssrn.2701902.

Ellen, I. (2015). Housing low-income households: lessons from the sharing economy? *Housing Policy Debate*, 25, 783–784; doi.org/10.1080/10511482.2015.1042204.

Elsworth, S. (2016). Airbnb blamed for overcrowding in high-rise buildings. *Herald Sun*.

Etienne, S. (2016). Arizona's Governor Ducey signs SB 1350 into law, prohibiting the ban of short-term rentals. TechCrunch.

EY (2016). The upside of disruption: Megatrends shaping 2016 and beyond. Retrieved from: www.ey.com/Publication/vwLUAssets/EY-the-upside-of-disruption/$FILE/EY-the-upside-of-disruption.pdf.

Fermino, J. (2015). NYC will spend $10M to crack down on illegal hotels. *New York Daily News*.

Fisher, T. (2015). Welcome to the Third Industrial Revolution: The mass-customisation of architecture, practice and education. *Architectural Design*, 8, 40–45; doi.org/10.1002/ad.1923.

Florence, P. (1948). *Investment, Location, and Size of Plant*. Cambridge University Press.

Gopal, P. and Perlberg, H. (2015). Airbnb hosts getting rich while traditional landlords lose out. Domain.

Greater London Council Deregulation Act (2015), s44; www.legislation.gov.uk/ukpga/2015/20/section/44/enacted.

Greater London Council (General Powers) Act (1973); www.legislation.gov.uk/ukpga/2015/20/notes/division/5/46.

Green, D. (2016). Hedge funder is fired after massive party that got him banned from Airbnb. *Business Insider Australia*.

Gurran, N. and Phibbs, P. (2017). When tourists move in: how should urban planners respond to Airbnb? *Journal of the American Planning Association*, 83(1): 80–92.

Guttentag, D. (2013). Airbnb: disruptive innovation and the rise of an informal tourism accommodation sector. *Current Issues in Tourism*, 15, 1192–1217; doi.org/10.1080/13683500.2013.827159.

Hall, P. and Florence, P. (1962). *Investment, Location, and Size of Plant*. Hutchison University Library.

Han, M. (2016). Airbnb managers take on real estate agencies. *Financial Review*.

Harrison, P. (2016). What I think every time I see an Airbnb renter in my neighborhood. Next City.

Heaton, A. (2016). Is Airbnb destroying strata living? Sourceable. Available from: https://sourceable.net/airbnb-destroying-strata-living/.

Hickey, S. and Cookney, F. (2016). Airbnb faces worldwide opposition. It plans a movement to rise up in its defence. *The Guardian*.

Hill, S. (2015). The unsavory side of Airbnb. *The American Prospect*.

Hunt, E. (2016). Airbnb a solution to middle-class inequality, company says. *The Guardian*. Retrieved from: www.theguardian.com/technology/2016/dec/14/airbnb-a-solution-to-middle-class-inequality-company-says.

Jefferson-Jones, J. (2015). Airbnb and the housing segment of the modern "sharing economy": Are short-term rental restrictions an unconstitutional taking? *Hastings Constitutional Law Quarterly*, 42.

Khadem, N. and Fuary-Wagner, I. (2016). Airbnb hosts should prepare for close scrutiny from the tax office, warns industry leader. Domain.

Krueger, L. (2016). Statement from Senator Krueger on Airbnb's misleading data. The New York State Senate; www.nysenate.gov/newsroom/press-releases/liz-krueger/statement-senator-krueger-airbnbs-misleading-data.

Kusisto, L. (2015). Airbnb pushes up apartment rents slightly, study says. *The Wall Street Journal*.

Lane, J. and Wooworth, R. (2016). The sharing economy checks in: an analysis of Airbnb in the United States. CBRE Hotels' Americas Research.

Lawler, R. (2012). Airbnb – Our guests stay longer and spend more than hotel guests, contributing $56M to the San Francisco Economy. TechCrunch.

Lee, T. (2017). Fully driverless cars could be months away. Ars Technica.

Lehane, C. (2016). Overhaul the laws for home-sharing. *New York Daily News*.

Lines, G. (2015). Hej, Not Hej Då: Regulating Airbnb in the new age of Arizona vacation rentals. *Arizona Law Review*, 57.

Loizos, C. (2017). As Uber's value slips on the secondary market, Lyft's is rising. TechCrunch.

Lunden, I. (2017). Airbnb closes $1B round at $31B valuation, profitable as of 2H 2016, no plans for IPO. TechCrunch.

Lynn, G. and Allen, A. (2017). Airbnb time limits "ineffective in London" councils say. BBC News.

Lyons, K. and Wearing, S. (2015). *The Sharing Economy: Issues, Impacts, and Regulatory Responses in the Context of the NSW Visitor Economy*. NSW Business Chamber, Sydney.

McCarthy, N. (2016). Which cities have the most Airbnb listings? Forbes.

Megginson, S. (2015). Supercharge your rental returns with short-term tenants. *Your Investment Property*.

Miller, S. (2016). First principles for regulating the sharing economy. *Harvard Journal on Legislation*, 53, 147.

Minifie, J. and Wiltshire, T. (2016). Peer-to-peer pressure policy for the sharing economy. *Grattan Institute Report*, 2016–7. Grattan Institute.

Mitchell, A. (2005). *The ESRI Guide to GIS Analysis*. ESRI Press.

Nemcova v Fairfield Rents Ltd (2016). UKUT 303 (LC) (6 September 2016); www.blasermills.co.uk/to-airbnb-or-not-to-airbnb-nemcova-v-fairfield-rents-ltd-2016-ukut-303-lc/.

New York State Office of the Attorney General (2014). *Airbnb in the City*. New York State Office of the Attorney General, New York.

NSW Finance Services and Innovation (2016). The Collaborative Economy in NSW – position paper.

NSW Parliament (2016). Adequacy of the regulation of short-term holiday letting in New South Wales; www.parliament.nsw.gov.au/committees/inquiries/Pages/inquiry-details.aspx?activetab=Reportsandpk=1956.

O'Donoghue, D. and Gleave, B. (2004). A note on methods for measuring industrial agglomeration. *Regional Studies*, 419–427; doi.org/10.1080/03434002000213932.

O'Neill, J. and Ouyang, Y. (2016). From air mattresses to unregulated business: an analysis of the other side of Airbnb. American Hotel and Lodging Association.

Orgill, M. (2015). Submission No 207 – Adequacy of the Regulation of Short-term Holiday Letting in New South Wales.

O'Sullivan, F. (n.d.). Europe's crackdown on Airbnb. CityLab.

Ozcan, P. (2016). Why Airbnb is welcome in some cities and not in others. The Conversation.

Quattrone, G., Proserpio, D., Quercia, D., Capra, L. and Musolesi, M. (2016). Who Benefits from the "Sharing" Economy of Airbnb, in: *Proceedings of the 26th International ACM Conference on World Wide Web 2–16*. Presented at the 26th International ACM Conference on World Wide Web 2016, Perth.

Reagor, C. (2017). Phoenix-area home sales on track for best year ever. *The Republic*.

Said, C. (2015). Airbnb's impact in San Francisco. *San Francisco Chronicle*.

Samaan, R. (2015). Airbnb, Rising rent, and the Housing Crisis in Los Angeles. LAANE.

Shatford, S. (2015). Hard data on the biggest Airbnb cities in the world. AirDNA.

Smith, A. (2016). Shared, collaborative and on demand: the new digital economy (No. 202.419.4372). Pew Research Center.

State of Arizona (2016). Senate Bill 1350, SB 1350.

Thomas, L. (2017). Airbnb just closed a $1billion and became profitable in 2016. CNBC.

Ting, I. (2016). How Aribnb is taking over Sydney, one beach at a time. *Sydney Morning Herald*.

Totten, S. (2017). By the numbers: Airbnb saw business blast this summer. *Phoenix Business Journal*.

US Census Bureau (n.d.). United States Census. Retrieved from www.census.gov/.

Valuation Office Agency (2017). Private rental market summary statistics: April 2016 to March 2017.

Wachsmuth, D. (2017). Airbnb and gentrification in New York. David Wachsmuth.

Wachsmuth, D. and Weisler, A. (2017). Airbnb and the Rent Gap: Gentrification Through the Sharing Economy. David Wachsmuth.

Wang, S. (2017). SoFi Raises $500 Million Led by Silver lake for Global Expansion. Bloomberg.

Zervas, G., Proserpio, D. and Byers, J. (2016). The rise of the sharing economy: estimating the impact of Airbnb on the hotel industry. Boston University School of Management Research Paper No. 2013–16; doi.org/10.2139/zsrn.2366898.

4

THE CONTRIBUTION OF GIS TO UNDERSTANDING RETAIL PROPERTY

Matthew Reed, Robert Buckmaster and Richard Reed

Introduction

This chapter examines 'retail destinations', a term which encompasses enclosed shopping centres, regional shopping centres, large format retail centres and bulky goods centres, as well as other forms of retail centres including high street and main street shopping precincts. In a similar manner to other spaces, a common scenario is a retail destination that is actually a mixture of land uses (e.g. retail centre with some office space; office building with some retail space) rather than a 100% allocation of retail space. Nevertheless, the research undertaken in this chapter regarding the application of GIS has implications for the different forms of retail land use, but is specifically designed for larger retail holdings.

In most competitive retail environments there are a range of different types of shopping options, in the form of retail destinations, which can quickly become out-of-date for consumers; accordingly, these options and strategies must be continually updated and implemented to ensure that increased customer sales and long-term traffic growth for retail destinations are sustained (Holland *et al.* 2012). It is envisaged there are both positive and negative attributes, which in turn would either positively or negatively affect the attractiveness of a retail destination from a consumer's perspective. The methodology will build on previous studies into retail destinations and customer attributes from the property/real estate, economic and mapping disciplines.

Literature review

Retail destinations

Previous studies identified clear linkages between (a) the delivery of retail services and (b) positive retail customer attraction and retention (Wong *et al.* 2001; Sit *et al.* 2003; Newman *et al.* 2007; Reimers *et al.* 2009; Tsai 2010; Kursunluoglu 2014). However, there remained uncertainty about which attributes directly contribute to customer attraction since research commenced into retail destinations and centres. For example, an early assumption was that, other than the actual location of the land itself, many retail destinations were homogenous and only offered variations in their size and travel distance (i.e. distance between the centre and consumer) metrics as the two main attributes of customer attraction (Huff 1962). However, later research confirmed that customers actively perceived a large range of different attributes associated with each retail destination as competitive differences (Nevin *et al.* 1980; Hise *et al.* 1983; Wong *et al.* 2001). It is now widely accepted that undertaking and quantifying customer demand is directly linked to the success of a retail destination; therefore substantial resources are usually allocated to ensuring that the design and functional relevance of a centre are aligned with customer demand (Teller *et al.* 2012).

Most stakeholders in the property and real estate industry acknowledge that retail destinations are no longer simply about purchasing products, but also involve an ancillary experience and associated enjoyment received by customers (Wong *et al.* 2001). There are varying terms used to describe a retail area or a precinct which includes retailers. For the purposes of this research the term 'retail destination' has been adopted, which includes reference to 'shopping centre', 'retail centre', 'shops', 'large format retail'. Therefore 'retail destination' is broadly relevant for a specific location where goods and services are being offered by a trader and where consumers visit. Often in property circles the term 'retail destination' is used interchangeably with 'shopping centre', which have a relatively high profile; however a 'shopping centre' is only one form of retail destination. In this context a shopping centre is often described as *'a group of retail and other commercial establishments that is planned, developed, owned or managed as a single property, comprised of multi-branded rental units and common areas'* (ICSC 2015).

The value drivers behind retail property are difficult to define for each retail destination since they are not as homogeneous as they may appear (API 2015). Notably the primary measure of success for a retail destination is directly linked to the level of customer expenditure on retail goods and services on a regular basis. Earlier landmark studies argued that retail destinations were homogenous, *'with the exception of size and distance'* from consumers; therefore only those two basic metrics were acknowledged as the main attributes of centre attractiveness (Huff 1962). However, subsequent studies confirmed that customers actively perceived competitive differences in retail destinations beyond their size. This shift acknowledged the impact of locational factors but recognised the increased importance of non-locational factors, including ambience and entertainment (Nevin *et al.* 1980; Hise *et al.* 1983).

The task of identifying and examining attributes of retail destinations is critical in today's competitive market-place, due to the need to provide both customer satisfaction and strong loyalty to differentiate the centre from other nearby competing retail destinations and to build a competitive advantage (Kursunluoglu 2014; Parsons and Ballantine 2004). An added complication affecting the level of consumer spend in retail destinations is the increased retail competition from online retailers such as Amazon; it is accepted that online retailers are consistently gaining market share, and offline retailers are seeking to arrest the decline while growing their own market share (Light *et al.* 2004). This trend is directly linked to the relative ease of home delivery for purchasers, again highlighting the importance of convenience for consumers. It should be noted that the retail industry is quite resilient; for example, most retail destinations did not suffer the predicted severe downturn and associated high level of obsolescence as predicted in the late 1990s with the widespread introduction and availability of the internet (Light *et al.* 2004). In contrast, retail destinations evolved to provide additional services, such as multi-faceted retail precincts with a diverse retail mix, cinemas, entertainment precincts and also food courts which collectively increased sales growth and produced an associated higher yield on investment.

FIGURE 4.1 Retail destination value drivers
Source: authors.

In recent times the retail market has diversified to include specific retailers with multiple tenancies in different retail destinations. For example, the shopping centre industry has been driven primarily by the real estate location rather than retailers, since the retail investment market continues to grow in both aggregate value and transaction activity (Holland *et al.* 2012). The value of a retail destination is strongly correlated with its net operating income; therefore the turnover and sales achieved by a retail destination are directly accountable, as illustrated in Figure 4.1. In addition the value drivers as highlighted in Figure 4.1 are circular, where each driver is equally important to ensure the success of a retail destination. For example, at the bottom of Figure 4.1 a new retail destination commences trading following initial capital investment in both land and structure, then is supported with marketing expenditure and raising the profile of the retail destination to the broader consumer market. When this step is undertaken effectively, it will be followed by increased foot traffic which will convert into retail sales as based on consumer spend – refer to the left-hand side of Figure 4.1. In order to increase the value of the retail destination and recover the initial capital investment/marketing costs, the shopping centre owner would most likely increase the level of rent, often linked to the retail turnover of a tenant – refer to the right-hand side of Figure 4.1. However, this is then followed by additional capital investment to ensure the retail destination remains competitive with other new and refurbished retail centres.

Characteristics of retail destinations

The retail market-place makes a substantial contribution to the global economy, including the provision of large scale employment. From an international perspective the revenue in the retail sector was estimated at US$22.6 trillion in 2015 and is predicted to rise to US$28 trillion by 2019 (Business Wire 2016); it was also confirmed that this sector represented 31% of the world's GDP and employed billions of workers, where hypermarkets and supermarkets accounted for 35% of direct retail sales. The total retail sales in the US for 2016 equated to US$5,484 billion, as shown in Table 4.1. Also highlighted is a breakdown of different categories which are typically included with reference to retail destinations. Note the size of the 'Total e-commerce retail sales' at US$384 billion in 2016, which equated to approximately 7% of all retail sales.

TABLE 4.1 Output and types of retail destinations in the US

Category	Amount (US$ billions)
Total retail sales in 2016	5,484.9
Total e-commerce retail sales in 2016	384.9
Motor vehicle and parts dealers	1,087.6
Furniture and home furnishings	105.4
Electronics and appliance stores	104.5
Building materials and garden equipment and supplies	332.6
Food and beverage stores	690.2
Health and personal care stores	312.8
Gasoline stations	432.0
Clothing and accessories stores	254.9
Sporting goods, hobby, book and music stores	90.4
General merchandise stores	673.6
Miscellaneous store retailers	121.1
Non-store retailers	503.2
Food services and drinking places	621.7
Total exports of goods	1,502.6
Total imports of goods	2,248.2
Employment in retail trade (x 1,000)	15,839.4
Number of US shopping centres	116,600

Source: Plunkett Research (2016).

A summary of information relating to the UK market in 2016 (Retail Economics 2016) is as follows:

- total value of UK retail sales was UK£358 billion with 2.8 million employees;
- 290,315 retail outlets existed in the UK;
- one-third of all UK consumer spending goes through retail;
- retail generated approximately 5% of total GDP in the UK;
- 12% of UK retail sales are made online, increasing by 10% annually;
- UK retail sales increase by 3.4% annually, with health and beauty increasing by 5.7% annually.

With reference to the retail market in Australia in 2016, there were 79,516 retail businesses equating to a total revenue of A$161 billion employing 734,301 individuals (Ibisworld 2016).

Locational attributes of retail destinations

While it has been established that locational attributes of customers are linked with the success of a retail destination, it is insightful and advantageous to use mapping to better understand the relationship with the location of the property. Traditionally GIS and mapping technologies have been under-utilised in the analysis of retail property and trends; however, exceptions exist with some fundamental and basic approaches (Roig-Tierno et al. 2013). Due to the global nature of retail destinations there is potential to increase the uptake of GIS analysis globally and benefit from the insights (Mishra and Nagar 2009).

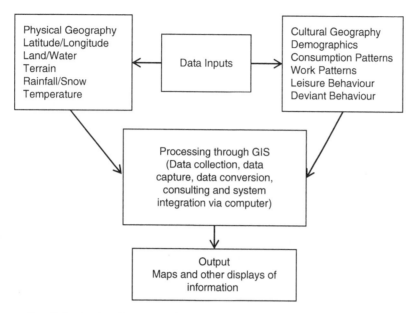

FIGURE 4.2 Retail destination drivers in a locational analysis
Source: based on Mishra and Nagar (2009).

The process for using mapping in a locational analysis for retail destinations is shown in Figure 4.2. The first step is to identify data inputs, which have been separated into the following two categories: (a) the left-hand side refers to physical and climatic variables and (b) the right-hand side refers to variables predominantly relating to consumer characteristics and behaviour. Note that these categories/variables are readily applicable to every retail destination; being consistent is an important consideration and requires careful identification of the variables to ensure they will contribute to the analysis and therefore not omitting any important variables from the inputs is key. The step following data collection is to enter these variables into a GIS analysis tool, and then undertake the tasks of data assembly, data capture, data conversion to a useable format, consultation and system integration into the mapping program. The final stage in the analysis is to produce an output in the form of maps (e.g. thematic maps) and other relevant displays of information.

Consumer behaviour – retail destinations

As highlighted in the approach to the locational analysis, the selection of the appropriate variables in the input phase will directly affect the effectiveness of the output. Whilst the physical locational variables (e.g. site area, proximity, building type) are fixed, there are variables relating to retail consumers which are relevant but more challenging to identify. The underlying framework for consumers visiting a retail destination is based on the long-standing 'central place' theory, where consumers usually visit the retail destination located in closest proximity when most other variables remain similar (Christaller 1933). Note that there are exceptions to this theory; for example, there can be ambiguous loyalty for a particular retail destination which is outside the scope of GIS analysis. This would occur where the individual human reasoning for these decisions is practically impossible to quantify; an example is a retail

TABLE 4.2 Examples of shopper categories

Type of retail consumer	Aim of shopper	Most common demographic cohort
Involved shoppers	Increase both utilitarian and hedonic shopping benefits	Female baby boomers, older shoppers
Experiential shoppers	Desire hedonic benefits	Female Generation X, Generation Y
Pragmatic shoppers	Increase utilitarian benefits and reduce shopping costs	Older male, Generation X, Generation Y
Nonchalant shoppers	Lowest level of concern of all shoppers	Male baby boomers

Source: Reed et al. (2005).

customer driving a large distance and bypassing existing retail destinations during transit. In a study examining consumer loyalty it was concluded that it is possible to gain loyalty from shopping centre customers travelling longer distances if a smooth relationship is established and maintained (Rabbanee et al. 2012). Therefore the GIS analysis will focus on central place theory as the underlying conceptual foundation, where proximity to the retail destination or ease of access is the starting point.

Prior to collecting the data inputs, the initial stage is to determine the target market for a specific retail destination based on relevant demographic cohorts. Many aspects relating to purchasing decisions can be linked to specific cohorts; for example, it has been shown that house purchasing decisions are closely associated with individual cohorts such as households with high loading socio-economic characteristics (Reed 2016). As shown in Table 4.2, it is essential to identify both (a) the type of retail consumer and (b) the most appropriate demographic cohort prior to collecting the data inputs. The skills associated with this decision are usually associated with managers of retail destinations and leasing experts, who are familiar with the breakdown of typical consumers based on their experience with previous retail destinations and their knowledge of competing centres and the broader retail market-place. Accurately identifying these inputs is not commonly undertaken in isolation by a mapping or GIS expert.

Listed in Table 4.3 is an example of the type of demographic variables which contributed to an analysis of consumer demand for a shopping centre in Wuhan, China (Rahman et al. 2017). The limitations of the study highlighted the relatively narrow focus of each shopping centre with reference to the relevant demographic variables; for example, the findings suggested the study should be replicated using different shopping malls and age groups, varying geographic locations and a wider range of target shoppers. This study also emphasised challenges when identifying relevant variables and inputs for a GIS study which often vary considerably between retail destinations.

Research question discussion

Owners of retail destinations continually seek to maximise both the asset value and the level of consumer attraction related to their centres, but clearly more emphasis needs to be placed on mapping consumer demand via GIS. Accordingly, the research question for this chapter is: *What are the drivers linked to retail customers visiting a specific retail destination rather than a competing destination?*

TABLE 4.3 Sample demographic variables in a retail consumer study

Variable	Shopper options
Sex	Male; female
Age	18–25; 26–30; 31–35; 36–40; 41–45; 46 or older
Marital status	Married; single; other
Number of children in household	0; 1; 2; 3 or more
Education	Junior high school; senior high school; college; undergraduate; graduate; others
Employment status	Employed, student, seeking employment, stay-at-home parent, retired
Monthly income (A$)	1,000 or lower; 1,001–2,000, 2,001–3,000, 3,001–4,000, 4,001–5,000, 5,001 and over

Source: authors.

Addressing this research question seeks to provide insight into the process for identifying the relevant drivers in order for the retail destination to be successful. Since their inception, most retail destinations have changed substantially from the original purpose for which they were established, being to provide the right product in a convenient location for customers to access (Ibrahim 2002). Based on fundamental economic theory that the number of suppliers will increase if there is demand and perceived profitability, the sector relating to retail has rapidly evolved into an extremely profitable and highly competitive environment in the 21st century (Holland *et al.* 2012).

A major challenge for retail destination owners is how to establish a stringent list of deliverables to present a strong offer to the consumer whilst also minimising exposure to risk for retail investors (Goodman *et al.* 2012). Within this context, attracting and growing the level of 'consumer retail spend' is the basic measure of the success of a retail destination such as a shopping centre or a large format retail centre (API 2015). However, the emphasis is clearly twofold, being (a) to attract a customer for their initial visit to the particular retail destination, and (b) attracting a customer repeatedly back to the same destination on a regular basis.

Research methodolody

This section conducts an analysis based on a case study approach. It uses the application of GIS to assess the potential for a retail location of a new store opening in Tasmania, the most southerly state in Australia.

Background to the case study

A case study is used to highlight the relevance of the mapping techniques to a 'real-life' scenario. The closure and decommissioning of the Burnie (Tasmania) Paper Mill made available a large, well-exposed, redundant brownfield site (20.33ha), available for reuse and redevelopment. A leading national home improvement chain with an extensive national network, but lacking presence in north-west Tasmania, sought to evaluate the potential to develop a large format store on part of the land parcel in Launceston. Previously the chain's nearest store to the proposed development was located in Launceston's southern suburbs, approximately 100km via road, with a total travel time of approximately 90 minutes between Burnie and

Launceston. Exit surveys conducted in the chain's existing stores in Launceston and Hobart confirmed that customers were typically prepared to travel for up to 30 minutes to access this type of retail destination. Hence the proposed new site stood to increase patronage into new territory with negligible reduction (i.e. cannibalisation) of the existing store's sales.

Subject to a favourable market assessment, a case was also required to be made to persuade the local government council (i.e. Burnie City Council in its role as the local planning authority) to rezone the land, as the proposed use and development was prohibited under the current industrial zone settings. This scenario is based on an engagement by the retailer's head office to prepare a market assessment of the proposed development with the assistance of GIS. The scope of the report was as follows:

- Assessment of the site and its immediate context;
- Definition of the catchment (i.e. customer trade areas) from which the proposed development was anticipated to derive its patronage and turnover, divided into its constituent parts;
- A demographic profile of the trade area including population forecasts;
- Retail spending estimates and forecasts for the trade area and its components;
- Assessment of the competitive landscape;
- A strategic justification for rezoning; and
- An executive summary outlining the key findings and contentions.

Site and location analysis

The site is located in the city of Burnie, a port and urban centre situated on Emu Bay at the mouth of Emu River in north-west Tasmania. It is located on the Bass Highway, which provides a direct connection to Launceston, located 148km to the east. With reference to the proposed land use, large format retailers typically require a relatively large site for the handling, display or storage, and direct vehicle access for the loading of goods after their purchase. Situated on the north-east corner of Marine Terrace and Reeve Street in South Burnie, this site enjoyed good access and extensive highway exposure. Note that Marine Terrace forms part of Bass Highway, where convenient vehicle access to the site was available from Reeve Street. With an area of approximately 4ha, the site had the capacity to accommodate the retailer's preferred large format, single-level store, together with an adequate amount of remaining area to facilitate customer parking.

The proposed site was located comparatively close to Burnie's central business district, being in easy walking distance. The precinct surrounding the site comprised a mix of light industrial, peripheral sales and residential development, where the industrial and commercial uses followed the low-lying land located on the southern side of the Bass Highway. Peripheral land uses associated with nearby property included used car yards and electrical supplies, with an electronic retail centre located approximately 700m north-west of the subject land towards the southern end of Burnie's central business district. At the commencement of the analysis the subject land was zoned as 'Industry'. The site itself was predominantly void of structural improvements (i.e. largely vacant) and under a single ownership, which afforded a marked advantage over other appropriately zoned alternative potential sites in Burnie. For example, the other sites were typically characterised by fragmented landownership necessitating costly and time-consuming site assembly from multiple landowners, potentially compromising the feasibility of any development.

The identification of the catchment area of a proposed retail destination is a critical step in quantifying the likely turnover, and provides a framework for identifying complementary and competitive land uses. The catchment or main trade area is the geographic area from which a retail destination derives most of its customers. The extent of this trade area is affected by a number of factors, including distance and travel times, natural features, built barriers, proximity of directly competitive alternatives, population density and neighbourhood boundaries and perceptions. When determining the extent of the main trade area the analysis had regard to the anticipated distance prospective customers were prepared to travel, informed by the patterns demonstrated by patrons in other parts of Tasmania together with the location of directly comparable competitors. A form of gravity modelling was undertaken, using a geographic information system to identify the breakpoint for the attraction of trade between the subject site in South Burnie and alternative retail destinations, most notably nearby Devonport.

With a broad and deep retail offering of higher order comparison goods, Burnie has traditionally functioned as a regional retail destination for the north-west corner of Tasmania, drawing customers from as far afield as Queenstown and Zeehan. The proposed development would service a main trade area extending to encompass the north-western corner of Tasmania. As shown in Figure 4.3, the primary trade area from which the development was anticipated to derive between 60 and 80% of its patronage generally coincides with the Burnie municipal district. This area is flanked by secondary east and secondary west catchments, respectively encompassing the towns of Penguin and Ulverstone, and Wynyard. A tertiary catchment area skirts the west of the secondary trade areas, encompassing the municipal districts of Circular Head (Stanley) and the West Coast (Queenstown/Zeehan).

Demographic profile of main trade area

A demographic profile was undertaken of the main trade area and constituent parts where the scope included:

- Existing and forecast population by constituent part over three-, five- and ten-year time horizons;
- Average annual population growth rate in percentage and absolute numbers; and
- An estimate of retail spending within the main trade area by its constituent parts by retail category, with a particular focus on home improvement merchandise.

The main trade area had an estimated resident population (ERP) of 59,600 in the middle of 2015. The area had experienced very modest population growth in recent years and this was anticipated to continue over the forecast horizon, with the main trade area population increasing to 60,350 residents by 2028. Furthermore, the trade area growth was anticipated to be focused within the primary trade area, with population loss anticipated in the tertiary trade area. The primary trade area resident population was projected to grow at an annual 0.2% (i.e. approximately 62 persons per annum) over the forecast horizon, increasing to approximately 26,439 residents by 2028. The main trade area was subdivided into primary, secondary and tertiary trade areas (see Table 4.4). Large format retail caters to a regional catchment as opposed to supermarkets, which cater to daily and weekly needs, tending to draw customers from a more localised area.

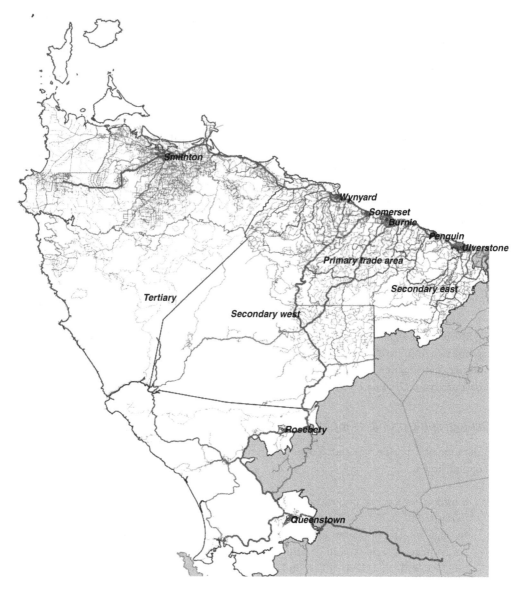

FIGURE 4.3 Map of main trade areas of proposed development
Source: authors.

Trade forecasts

Per capital retail spending estimates were prepared, based on data sourced from the Australian Bureau of Statistics (ABS), which publishes monthly retail trade data, the ABS *Household Expenditure Survey* (ABS 2015) and the trade area's income and demographic profile. The ABS publishes retail trade (i.e. turnover) figures monthly, disaggregated by state and major retailer categories. Household expenditure data is collected every five years and disaggregated by individual products, services and income brackets. For this analysis it was estimated that, in the primary trade area, each resident spent an average A\$2,219 per annum on large format retail

TABLE 4.4 Main trade area population (2011–2028)

Population per trade area	2011	2015	2018	2021	2023	2028
Primary trade area	25 896	25 636	25 819	26 004	26 127	26 439
Secondary trade area						
Secondary east	11 270	11 320	11 494	11 671	11 790	12 095
Secondary west	9 801	9 857	9 857	9 857	9 857	9 857
Total secondary trade area	21 071	21 176	21 351	21 528	21 647	21 951
Tertiary trade area	13 251	12 764	12 573	12 386	12 262	11 959
Main trade area	60 218	59 576	59 743	59 917	60 037	60 350
Average annual growth (%)						
Primary trade area		-0.3%	0.2%	0.2%	0.2%	0.2%
Secondary trade area						
Secondary east		0.1%	0.5%	0.5%	0.5%	0.5%
Secondary west		0.1%	0.0%	0.0%	0.0%	0.0%
Total secondary trade area		0.1%	0.3%	0.3%	0.3%	0.3%
Tertiary trade area		-0.9%	-0.5%	-0.5%	-0.5%	-0.5%
Main trade area		-0.3%	0.1%	0.1%	0.1%	0.1%
Average annual growth (no.)						
Primary trade area		(65)	61	62	62	62
Secondary trade area						
Secondary east		12	58	59	60	61
Secondary west		14	0	0	0	0
Total secondary trade area		26	58	59	60	61
Tertiary trade area		(122)	(64)	(63)	(62)	(61)
Main trade area		(160)	56	58	60	63

Source: Australian Bureau of Statistics (2017a), Tasmanian Government (2016), Location Choice Strategists (2017).

goods in 2015. This amount ranged from between A\$2,188 to A\$2,245 across the balance of the main trade area (see Table 4.5). Estimates of aggregate retail spending by category in the main trade area and its constituent parts were modelled, based on data relating to household income and demographic characteristics. In this analysis it was estimated that A\$132.0 million would be spent on large format goods across the main trade area in 2015, increasing to approximately A\$203.9 million by 2028. Within the primary trade area the expenditure estimates were A\$56.9 million and A\$89.5 million respectively (see Table 4.6).

Competitive landscape assessment

The assessment also involved identifying existing competitors serving the catchment and estimating their turnover. Existing home improvement stores were identified via a mainstream internet search engine and a telephone directory search, together with ground reconnaissance. The next step was to use a web-based aerial photography service ('Nearmap') to estimate the selling area from the aerial footprint of these competitors. Then the amount of turnover was estimated, based on application of industry turnover benchmark averages and observations from field trips. It was then determined that a total of 17 large format retailers with an aggregate floor area of 28,100m² operated within the primary trade area. An additional 17 stores

TABLE 4.5 Average per capita retail spending and projected annual nominal growth (A$)

	Primary trade Area	Secondary trade Area East	Secondary trade Area West	Tertiary trade Area	Forecast annual Increase
Supermarket and grocery stores	$4,074	$4,020	$4,048	$4,143	4.1%
Liquor stores	$349	$343	$347	$354	4.4%
Other food	$275	$271	$273	$279	4.1%
Pharmaceutical, cosmetic and toiletry goods retailing	$585	$575	$578	$599	3.9%
Food and liquor retailing	$5,283	$5,209	$5,246	$5,377	
Furniture, floor coverings and other goods wholesaling	$568	$562	$559	$574	2.8%
Electrical and electronic goods	$666	$659	$655	$673	2.9%
Hardware, building and garden supplies	$795	$787	$783	$804	3.9%
Other recreational goods	$191	$189	$190	$194	3.5%
Large format retail goods	$2,219	$296	$288	$2,245	3.3%

Source: Australian Bureau of Statistics (2017b), Location Choice Strategists (2017).

with an aggregate area of 18,799m² located in the balance of the main trade area were also identified.

Results – case for development approval

Since the proposed use and development was prohibited under the existing industrial zoning, the realisation of the concept was dependent on persuading the local council to rezone the subject land and permit a different land use. Accordingly, this required a strategic justification for a site-specific application to enable an amendment to the planning scheme. The final report contended that the development of large format retail at this location was consistent with Burnie Council's adopted strategic framework for the city's future development and growth, and also with the findings of the Burnie Paper Mill Site Assessment Report (AEC 2010). This report, as commissioned by Burnie Council, identified the highest and best uses of the redundant mill site. It was further contended that large format retail was a suitable use with regard to surrounding development, including existing peripheral sales, the quasi-industrial profile of the proposed use and the limited availability of alternative sites considered fit for purpose elsewhere in the municipal district. The report supported use of the site as a logical extension of the South Burnie bulky goods/homemaker precinct. Brownfield redevelopment (i.e. redevelopment of a disused industrial site) was advocated as a viable means of assisting urban consolidation and renewal. In due course the Burnie Council unanimously resolved to amend the scheme and grant the permit for alternative land use.

TABLE 4.6 Estimated and projected main trade area retail spending

Retail spending (A$m)	2015	2018	2021	2023	2028
Primary trade area					
Furniture, floor coverings, houseware and textile goods retailing	14.6	15.9	17.5	18.6	21.6
Electrical and electronic goods retailing	17.1	18.7	20.5	21.8	25.5
Hardware, building and garden supplies retailing	20.4	23.0	26.0	28.2	34.5
Other recreational goods	4.9	5.5	6.1	6.6	7.9
Large format retail goods	*56.9*	*63.1*	*70.1*	*75.2*	*89.5*
Secondary trade area east					
Furniture, floor coverings, houseware and textile goods retailing	6.4	7.0	7.8	8.3	9.8
Electrical and electronic goods retailing	7.5	8.2	9.1	9.8	11.5
Hardware, building and garden supplies retailing	8.9	10.1	11.5	12.6	15.6
Other recreational goods	2.1	2.4	2.7	2.9	3.6
Large format retail goods	*24.9*	*27.8*	*31.1*	*33.6*	*40.5*
Secondary trade area west					
Furniture, floor coverings, houseware and textile goods retailing	5.5	6.0	6.5	6.9	7.9
Electrical and electronic goods retailing	6.5	7.0	7.7	8.1	9.4
Hardware, building and garden supplies retailing	7.7	8.7	9.7	10.5	12.7
Other recreational goods	1.9	2.1	2.3	2.5	2.9
Large format retail goods	*21.6*	*23.8*	*26.2*	*28.0*	*32.9*
Tertiary trade area					
Furniture, floor coverings, houseware and textile goods retailing	7.3	7.9	8.4	8.8	9.9
Electrical and electronic goods retailing	8.6	9.2	9.9	10.4	11.7
Hardware, building and garden supplies retailing	10.3	11.3	12.5	13.4	15.8
Other recreational goods	2.5	2.7	3.0	3.1	3.6
Large format retail goods	*28.7*	*31.1*	*33.8*	*35.7*	*41.0*
Main trade area					
Furniture, floor coverings, houseware and textile goods retailing	33.7	36.8	40.2	42.6	49.2
Electrical and electronic goods retailing	39.6	43.2	47.2	50.1	58.0
Hardware, building and garden supplies retailing	47.3	53.2	59.8	64.6	78.6
Other recreational goods	11.4	12.6	14.1	15.1	18.0
Large format retail goods	*132.0*	*145.8*	*161.2*	*172.4*	*203.9*

Source: Australian Bureau of Statistics (2017b), Location Choice Strategists (2017).

In response to the stated research question for this chapter, the drivers identified in this case study approach were as follows:

- Conducting a site and location analysis;
- Undertaking a trade forecast;
- Examining the demographic profile of the potential customer base;
- Undertaking a competitive landscape assessment including examining the location of competing retailers who supply similar consumer goods;

- Identifying the highest and best use of the subject site;
- Examining the relevant planning framework for the surrounding area including the subject site; and
- Analysing the trade areas based on primary trade area, secondary trade area and tertiary trade area.

Conclusion

This chapter examined the potential for GIS and mapping to contribute to the spatial analysis of a potential retail destination. It included an analysis of retail property, the contribution of GIS and also used a complex case study based on a potential new site for the location of a retail destination in Tasmania. The research findings confirmed there are multiple drivers from a GIS perspective and it is essential to integrate spatial analysis into this type of study. Omission of a detailed spatial analysis would adversely affect the reliability of the findings and the subsequent success or failure of the large scale development. Whilst this analysis included the traditional demographic variables relative to potential consumers, there remain many other factors which must also be considered. These include consideration of locational factors such as where existing or competing retail stores are located, and specific information relating to trade area, local planning considerations, as well as highest and best use of the subject site.

References

AEC (2010). *Burnie Paper Mill Site Assessment Report*. AEC Group: Melbourne.

API (2015). *The Valuation of Real Estate*. Australian Property Institute: Canberra.

Australian Bureau of Statistics (2015). 6530.0, *Household Expenditure Survey*. ABS: Canberra.

Australian Bureau of Statistics (2017a). 3218.0, *Regional Population Growth Australia* ABS: Canberra.

Australian Bureau of Statistics (2017b). 8501.0, *Retail Trade Australia*. ABS: Canberra.

Berger, R. (2015). Now I see it, now I don't: researcher's position and reflexivity in qualitative research. *Qualitative Research*, 15(2), 219–234.

Business Wire (2016). Global Retail Industry Worth USD 28 Trillion by 2019. Retrieved from: www.businesswire.com/news/home/20160630005551/en/Global-Retail-Industry-Worth-USD-28-Trillion.

Christaller, W. (1933). *Die Zentralen Orte in Suddeutschland, Fischer, Jena, 1966*; English translation from German original by L.W. Baskin, Central Places in Southern Germany. Prentice Hall: Englewood Cliffs, NJ.

Goodman, R. and Coiacetto, E. (2012). Shopping streets or malls: changes in retail form in Melbourne or Brisbane. *Urban Policy and Research*, 30(3), 251–273.

Hise, R., Kelly, J., Gable, M. and McDonald, J. (1983). Factors affecting the performance of individual chain store units: an empirical analysis. *Journal of Retailing*, 59(2), 22–39.

Holland, P. and Backen, J. (2012). Australia: Short term problems but long-term opportunities. *Retail Property Insights*, 19(2), 45–49.

Huff, D. (1962). Determination of intra-urban retail trade areas. *Real Estate Research Program*, University of California, 11–12.

Ibrahim, M. (2002). Disaggregating the travel components in shopping centre choice: An agenda for valuation practices. *Journal of Property Investment and Finance*, 20(3), 277–294.

IBISWorld (2016). Consumer goods retail market research report – Aug 2016. IBISworld. Retrieved from: www.ibisworld.com.au/industry/consumer-goods-retail.html.

ICSC (2015). Asia-Pacific shopping centre definition standard. International Council of Shopping Centres. Retrieved from: www.icsc.org/research/references/c-shopping-center-definitions.

5

MODELLING VALUE UPLIFT ON FUTURE TRANSPORT INFRASTRUCTURE

Scott N. Lieske, Ryan van den Nouwelant,
Hoon Han and Chris Pettit

Introduction

There is a renewed interest in 'value capture' funding for infrastructure in Australian metropolitan planning. There is a direct financial cost of providing infrastructure and there is often a positive externality to landowners surrounding that infrastructure. This nexus has long been a justification for a taxation structure that taps into the latter to fund the former. In the current era of tighter government budgets, this source of funds has become increasingly appealing.

However, quantifying the extent to which some infrastructure is capitalised in surrounding property values is an ongoing barrier to an effective but fair value capture mechanism. More importantly, such quantification takes time and resources so is often done after crucial infrastructure planning decisions are made. These are inherently spatial challenges and this chapter outlines how spatial technologies such as Geographical Information Systems (GIS) and Planning Support Systems (PSS) can rapidly estimate the effect of infrastructure on property values. In turn, this can inform decision-makers about the relationship between infrastructure provision and changing property values.

The chapter describes the creation and application of a rapid spatial analysis decision-support platform using a case study of train infrastructure in the western suburbs of Sydney, Australia. The platform (a) organises diverse data sets into a coherent, property level database, then (b) uses that database to generate a hedonic price model, determining the impact of transport infrastructure on property prices, and (c) estimates the value uplift generated by a hypothetical train line based on the hedonic price model.

Project importance: a renewed interest in value capture

Australia's transport infrastructure planning is under pressure on two fronts. First, there is increasing pressure to make up for a legacy of under-investment in public transport infrastructure in suburban locations. Rapid population increase is concentrated in the capital cities (McGuirk and Argent 2011; Elaurant and Louise 2015; ABS 2017), and there is historically lower transportation access in the middle and outer ring suburbs which are taking on much of this new population growth (Randolph and Tice 2014). The urgency was expressed by the NSW State Minister for Transport Andrew Constance when describing a rail line through the western suburbs of Sydney:

Kursunluoglu, E. (2014). Shopping centre customer service: creating customer satisfaction and loyalty. *Marketing Intelligence and Planning*, 32(4), 528–548.

Light, D. and Kenins, I. (2004). The shop of things to come. *Bulletin with Newsweek*, 122(6411), 34–37.

Location Choice Strategists (2017). Personal comment, R. Buckmaster.

Mishra, S. and Nagar, R. (2009). GIS in Indian retail industry – a strategic tool. *International Journal of Marketing Studies*, 1, 50–57.

Nevin, J. and Houston, M. (1980). Image as a component of attraction to intra-urban shopping areas. *Journal of Retailing*, 56(1), 77–93.

Newman, A., Dennis, C. and Zaman, S. (2007). Marketing images and consumers' experiences in selling environments. *Marketing Management Journal*, 17(1), 136–150.

Parsons, A. and Ballantine, P. (2004). Market dominance, promotions and shopping mall group performance. *International Journal of Retail and Distribution Management*, 32(10), 458–463.

Plunkett Research (2016). U.S. Retail Industry Statistics and Market Size Overview, Business and Industry. Retrieved from: www.plunkettresearch.com/statistics/Industry-Statistics-US-Retail-Industry-Statistics-and-Market-Size-Overview/.

Rabbanee, F.K., Ramaseshan, B., Wub, C. and Vinden, A. (2012). Effects of store loyalty on shopping mall loyalty. *Journal of Retailing and Consumer Services*, 19, 271–278.

Rahman, O., Fung, B.C.M., Chen, Z., Chang, W.-L. and Gao, X. (2017). A study of apparel consumer behaviour in China and Taiwan. *International Journal of Fashion Design, Technology and Education*, 11(1), 22–33; 10.1080/17543266.2017.1298158.

Reed, Richard (2016). The relationship between house prices and demographic variables: An Australian case study. *International Journal of Housing Markets and Analysis*, 9(4), 520–537; doi.org/10.1108/IJHMA-02-2016-0013.

Reed, R.G. and Conisbee, N. (2005). Identifying linkages between generations and community development: the effect on residential and retail property. In *Proceedings of the 11th Annual Pacific Rim Real Estate Society Conference*, Melbourne, 23–27/01/05.

Reimers, V. and Clulow, V. (2009). Retail centres: it's time to make them convenient. *International Journal of Retail and Distribution Management*, 37(7), 541–562.

Retail Economics (2016). UK retail stats and facts. Retrieved from: www.retaileconomics.co.uk/library-retail-stats-and-facts.asp. Date accessed 01/06/2017

Roig-Tierno, N., Baviera-Puig, A. and Buitrago-Vera, J. (2013). Business opportunities analysis using GIS: the retail distribution sector. *Global Business Perspective*, 1, 226–238.

Sit, J., Merrilees, B. and Birch, D. (2003). Entertainment-seeking shopping centre patrons: the missing segments. *International Journal of Retail and Distribution Management*, 31(2), 80–94.

Tasmanian Government (2016). Population projections for Tasmania and its Local Government Areas. Department of Treasury and Finance.

Teller, C., Kotzab, H. and Grant, D.B. (2012). Improving the execution of supply chain management in organizations. *International Journal of Production Economics*, 140(2), 713–720.

Tsai, Ming-Chih, Hsin, Wen Lee and Wu, Chieh (2010). Determinants of RFID adoption intention: Evidence from Taiwanese retail chains. *Journal of Information and Management*, 47(5–6), 255–261. Retrieved from: www.sciencedirect.com/science/article/abs/pii/S0378720610000480.

Wong, G., Yu, L. and Lim Lan, Y. (2001). SCATTR: an instrument for measuring shopping centre attractiveness. *International Journal of Retail and Distribution Management*, 29(2), 76.

Mr Constance said the new line was a "must build" because the existing T1 Western Line would reach full capacity by 2031, which "means quite literally you will not be able to get people physically onto the trains"

(O'Sullivan 2017).

Second, there is increasing pressure to pay for this infrastructure from outside traditional government budgets. Since the 1980s there has been a shift from government provision of infrastructure to government procurement of infrastructure (O'Neill 2010). This has been coupled with the dominance of neoliberal political rhetoric (Randolph and Tice 2017), stressing minimising taxation and so maximising 'user pays' in infrastructure funding. While this partly translates to operational revenues of the transport infrastructure, it also increasingly means capturing part of the increase in land values caused by the infrastructure provision.

Investment in public transport infrastructure benefits surrounding residents and businesses. New and improved transport – including new stations, faster trips, more frequent trips, upgraded services or higher volume services – can connect residents to jobs and services and also connect businesses to labour, supplier and customer markets (McIntosh *et al.* 2016). As such, there is an expectation that residents and businesses will pay a premium for property serviced by this infrastructure. That is, infrastructure investment will increase residential and commercial property values, as an increased willingness to pay for this improved service is capitalised into nearby property prices.

This anticipated increase in property value associated with transport infrastructure is called 'value uplift'. Government appropriation of a portion of that value uplift, to fund current or future transportation infrastructure projects, is called 'value capture' (or 'value sharing'). Value capture is, in essence, a tax on the increase in land values associated with new or upgraded infrastructure, and these taxes are used to fund the project (Terrill and Emslie 2017). Quantifying value *uplift* is a key step in evaluating the feasibility of value *capture* (Mulley 2014).

Value capture is currently widely used internationally to fund major infrastructure projects. For example, nearly one third of the cost of London's CrossRail (due to be completed in 2019) will be funded by a 30-year rates supplement to the largest 20% of commercial properties in Greater London (Terrill and Emslie 2017). Historically Australia has used value capture to fund transport infrastructure. For example, the Sydney Harbour Bridge, built in the 1920s, was partly funded through a 'betterment levy' paid by landowners in the local councils with improved access to the central business district (Terrill and Emslie 2017).

Terrill and Emslie (2017, p.3) summarised the renewed interest in value capture across all levels of government in Australia: 'Value capture is back in fashion, and the calls are growing louder for Australia to tap into these seemingly wonderful revenue streams.' State infrastructure plans increasingly support value capture (Terrill and Emslie 2017) and the Commonwealth Government's *Smart Cities Plan* (2016) stresses that value capture needs to be explored as a funding option in all cases where Commonwealth Government capital is sought. The *Smart Cities Plan* (2016) argues that value capture can make infrastructure more affordable, and speed up project delivery as well as increases in housing supply and urban renewal.

There are, though, known gaps in how value capture can be used in practice in current Australian projects. Highlighting the importance of this sort of information, the Australian Government has committed A$50 million to building the business cases and project delivery models for infrastructure projects financed with value capture (Turnbull 2016). Infrastructure Australia (2016) further highlighted the challenges, noting the difficulty of separating the influence of infrastructure from other factors that influence property values.

A recent notable example of value capture is the Gold Coast Light Rail, where construction began in 2010. However, as Terrill and Emslie (2017) indicated, the levy charged in that case was not geographically linked to access to the infrastructure or anticipated land value increases, making it less efficient. Murray (2016) estimated land value gains to nearby landowners associated with development of the Gold Coast Light Rail to be A$300 million. This value uplift is 25% of the capital cost of the project, demonstrating the potential to fund transportation infrastructure investment through smarter value capture mechanisms.

Value uplift and value capture are inherently spatial phenomena, so there is a clear potential to inform infrastructure decision-making with geographic information systems (GIS) and planning support systems (PSS). As outlined in the following sections, this chapter demonstrates how a rapid spatial analysis decision-support platform can increase the understanding of value uplift in infrastructure planning.

Project contribution: quantifying the impact of infrastructure on property values

This section outlines the development and implementation of a prototype toolkit, known as the Rapid Analytics Interactive Scenario Explorer (RAISE) toolkit. The toolkit uses hedonic price modelling (HPM) to quantify the impact of transport infrastructure on residential property prices, then uses this modelling to rapidly estimate changes in property values during the exploration of different hypothetical scenarios of new infrastructure provision. That is, by incorporating the database development, modelling and scenario exploration into a PSS, the project's contribution is a *method* for the rapid quantification of value uplift. In turn, this can inform the development of value capture mechanisms that can efficiently levy property value based on the contribution of the infrastructure to that value. The method is intended to be adaptable to use in other cases.

The prototype was developed for a case study of residential properties in the Parramatta City Council local government jurisdiction and surrounding suburbs. Parramatta is Sydney's third major employment centre (see Figure 5.1) and the case study focused on the value of commuter rail infrastructure. The study area for this research was based on the 2015 boundary of the Parramatta City Council, but extends to all lots in the circumcircle of the council boundary; this includes potentially relevant sales in close proximity to the council that would otherwise be excluded with strict adherence to the political boundary.

Hedonic price modelling is a commonly accepted and longstanding technique in land and property market analysis (Rosen 1974). It serves two primary functions. The first is to decompose market prices based on property characteristics, or attributes, identifying the marginal influence on price of each attribute independently. The second is to estimate property values, in the absence of a sale price, based on those property attributes.

Conceptually, HPM follows the idea that the value of goods bought and sold in a market are determined by some bundle of characteristics. For residential properties this includes (Costello and Watkins 2002):

- Structural attributes such as the building's quality, age and size;
- Neighbourhood attributes such as crime levels and socio-economic profile; and
- Accessibility attributes such as nearest transport connections, job centres and services.

Empirically HPM is a form of multivariate regression which uses a market indicator (e.g. sale price) as the dependent variable and some quantifiable measure of the attributes (as described

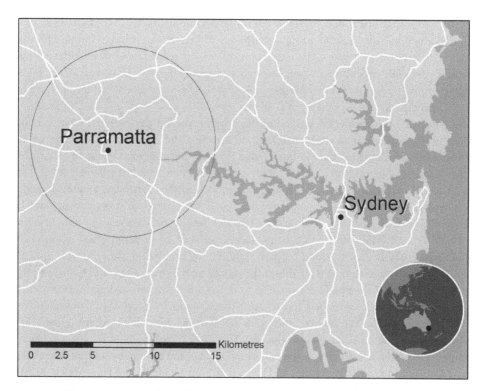

FIGURE 5.1 The Parramatta circle study area in Western Sydney, New South Wales, Australia
Source: authors.

above) as the independent variables. This enables the development of a formula that best estimates prices for properties by determining the weightings for the set of attributes across a number of sold properties (van Kooten 1993). That is, the modelling can determine the coefficients (C_0 to C_n in the generic formula) in equation 5.1, based on properties with known sales prices (V) and known attribute values (A_1 to A_n).

$$V = C_0 + (A_1 \times C_1) + (A_2 \times C_2) + (A_3 \times C_3) + \dots + (A_n \times C_n) \qquad [5.1]$$

Once the coefficients that best model the observed data are established, the formula can be used to estimate the value (V) for other properties, even in the absence of a market indicator, based on those coefficients and the known attributes of those properties. By identifying the marginal effect of each attribute, HPM also allows an estimate of the difference in property value if an attribute value changes. This is exploited in the RAISE toolkit, which, by introducing hypothetical train stations, changes the 'distance to nearest train station' attribute in the formula. Selecting appropriate attributes and appropriate quantifiable measures of those attributes is a key to developing an accurate model.

Hedonic analysis is often used to disaggregate housing prices by accessibility characteristics including access to commuter rail. Páez et al. (2012) defined accessibility as the potential for reaching nearby infrastructure from a particular location in association with travel cost or walkability. Accessibility is often estimated by distance (e.g. metres) from nearby infrastructure. Higgins and Kanaroglou (2017) used a HPM to estimate an impact of both rapid transit and associated transit orientated development on land value uplift. The HPM shows significant price premiums for locations around stations as well as variation in land value uplift by station context. For these reasons, the estimation of hedonic price functions includes a range of housing and built environment variables such as dwelling density (Han *et al.* 2017), high-voltage overhead transmission lines (Wadley *et al.* 2017), low-income rental housing (Davison *et al.* 2016) and public park/open space (Chhetri *et al.* 2009).

The objective of the project described in this chapter is to integrate the HPM into a GIS and PSS platform. There is an established method for using PSS as a research tool to support the development of information and knowledge for planning and governance of sustainable urban systems (see Geertman *et al.* 2017; Lieske *et al.* 2015; Pettit *et al.* 2015; Wadell 2002). This allows decision-makers to draw together data and models, rapidly formulate and test hypotheses, as well as receive immediate visual feedback (e.g. via interactive maps and graphs). In turn, this allows decision-makers to better understand how spatial data sets and models relate to each other. In RAISE, the toolkit helps decision-makers to understand the connection between infrastructure provision and property values using 'real world' data.

The RAISE prototype uses *ArcGIS* (ESRI Inc., Redlands, California, US) and *Community Viz* (City Explained Inc., Lafayette, Colorado, US) to develop the data sets and perform investigation into the data and related models. In particular, the Scenario 360 module of CommunityViz extends the quantitative capabilities of ArcGIS by allowing formula-based spreadsheet-like calculations to be performed on geographic data. This allows flexible adjustment of geographic and numeric inputs as well as automated recalculation of maps and quantitative output in a process referred to as 'dynamic analysis' (Walker and Daniels 2011, p.32; Lieske and Hamerlinck 2015).

The capabilities and value of PSS as a method for estimating the impact of infrastructure provision on property values are threefold:

1. The ability to collect, store and wrangle dynamic property, neighbourhood and environmental data into a format suitable for hedonic price modelling (Bastian *et al.* 2002);
2. The ability to connect those datasets with an econometric modelling package used to determine statistically significant determinants of price;
3. The ability to incorporate the calculated marginal influence of significant property characteristics (i.e. coefficients) within an interactive scenario explorative process, therefore enabling new transport infrastructure to be tested with real-time feedback given based on changes in value uplift.

Methodological details of each capability are presented in the next section.

Project development: PSS development and modelling

Setting up the decision-making framework

The first stage of analysis was to consolidate and aggregate data within a GIS database representing potential determinants of residential property price. Data were obtained from different sources at a variety of spatial resolutions. This section describes how the CommunityViz formula editor was used to construct a dataset of dynamic attributes. Dynamic attributes are crucial to scenario analysis where spatial attributes (e.g. the location of the nearest train station) are manipulated. Dynamic attributes are formula-based GIS data attributes (Walker and Daniels 2011) that update automatically to changes in numerical or spatial data inputs. Similar to, but considerably more detailed than the ArcGIS field calculator, the CommunityViz formula editor facilitates writing formulas that execute numeric and/or spatial calculations and offers the flexibility and much of the power of a computer programming environment without the need for specialist GIS coding skills.

The spatial basis of the analysis was the 'lots' layer of the NSW Digital Cadastral Database (DCDB). Thie (2008, p.15) defined a cadastre as 'a parcel-based and official geographical information system that renders identification and attributes of the parcels of a jurisdiction'. Cadastral data, including the lot, property or parcel boundaries, are the authoritative source of information on the identity, ownership and spatial extent of land. Advantages of cadastral data include continuous updating as well as best available locational and data attribute information. Most importantly, a land parcel is the physical unit of legal title and a core element used in decision-making about land use changes (Lieske and Gribb 2012).

A suite of different formula types was used for data aggregation, analysis, modelling and to facilitate user interaction. These included:

- *Get* formulas which draw values from other data layers into the current layer based on location or attributes;
- *Count* formulas which populate a dynamic attribute by tallying features from another layer based on location or attributes;
- *MinDistance* formulas which calculate direct or network distance between a feature in the lots layer and a feature in another layer, again depending on location or other attributes;
- *IfThenElse* formulas which evaluate a condition and return a specified value if the condition is true;
- *GetFromClosest* formulas which employ a spatial test to populate a dynamic attribute by drawing features from another layer based on attributes or an additional spatial test;
- *Coefficient-based* formulas which are a subset of numeric formulas that include rates or other modifiers that can be applied to numeric and/or spatial data (Walker and Daniels 2011); and
- *numeric* formulas which are straightforward mathematical calculations.

A *get* formula was used to transfer residential sales price data into the lots database. The dependent variable in the hedonic price model is sales price, 'EventPrice', sourced from the Australian Property Monitors (APM) Sydney Geocoded sales data. In order to minimise the influence of temporal trends in the property market, the scope of this analysis only includes sales data and associated structural attributes from the 2015 financial year. The 2015 dataset includes sales events recorded from 1 July 2014 to 30 June 2015. EventPrice is a sales price, measured in Australian dollars, associated with an event date (e.g. sale recording date) also indicated in the APM dataset. Original APM data were provided in tabular form with a latitude and longitude recorded for each sale, indicating location. APM data were added to the GIS as point data, clipped to the Parramatta circle study area and projected to Geocentric Datum of Australia (GDA94) Map Grid of Australia (MGA) UTM Zone 56.

Adding the EventPrice data to the lots layer was a two-step process. In order to address circumstances where there were multiple sales within the geographic confines of a single lot, for example due to multiple sales or sales of strata units, the APM 2015 data were iterated through an ArcGIS model builder script (see Figure 5.2), so a lot from the Property NSW cadastre was selected for each data record in APM 2015. The modelbuilder script also populated a field

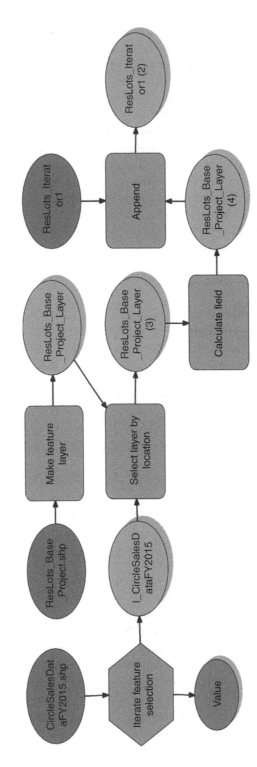

FIGURE 5.2 ArcGIS Modelbuilder script linking lots to sales data

Source: authors.

(AFID) which provides a one-to-one link between each lot and its corresponding record in the APM database. *ArcGIS model builder* offers a number of advantages over manual GIS processing, including reduction of time and ensuring consistent application of methods in repetitive and iterative processes.

A CommunityViz dynamic attribute was then used to calculate 'EventPrice', where point observations of sales data are linked to residential lots using feature ID values. EventPrice is created as a dynamic attribute in the lots database. Equation 5.2 is a *get ... where* formula that links the lot data with the APM Sydney Geocoded sales data:

$$\text{Get ([Attribute:CircleSalesDataFY2015:EventPrice],} \qquad [5.2]$$
$$\text{Where ([Attribute:CircleSalesDataFY2015:FID] = [Attribute:AFID]))}$$

Get ... where formulas are also employed to bring a number of other attributes of the APM Sydney Geocoded sales data into the lots data layer. Similar to EventPrice, both bedrooms and bathrooms data were brought into the residential lots layer using a CommunityViz dynamic attribute to obtain the data attribute from the APM 2015 sales data record by feature ID. With little data about building size, footprint or actual height, we follow Mulley (2014) who with similar data used the number of bedrooms, bathrooms and parking spaces to control for residence size in the hedonic model.

Count formulas populate a dynamic attribute by tallying features from another layer based on location or attributes (see equation 5.3). Numerous property characteristics from the APM Sydney Geocoded sales data were aggregated within the lots layer using a count formula, including walk-in wardrobe, rumpus room, internal laundry and balcony. Within the APM data these and many other property characteristics simply indicate presence or absence of a characteristic for each property. While the *count* formula is the most parsimonious manner of bringing these values into the lots layer, they are Boolean values indicated as one (true) or zero (false) and not actual counts.

$$\text{Count ([Attribute:CircleSalesDataFY2015:HasWalkInW],} \qquad [5.3]$$
$$\text{Where ([Attribute:CircleSalesDataFY2015:FID] = [Attribute:AFID] And}$$
$$\text{[Attribute:CircleSalesDataFY2015:HasWalkInW] = "TRUE"))}$$

A distance formula, *MinDistance*, was used to calculate the direct distance between a feature in the lots layer and a feature in another layer depending on location or other attributes. A number of dynamic attributes in the lots layer were calculated using a *MinDistance* formula that included an optional *where* clause; this included distance to train station, distance to different land use zones (e.g. commercial, industrial), distance to primary and secondary schools, distance to a community pool, distance to the children's hospital and distance to a highway on/off ramp. For example, Dist2_ZoneIND123 is the distance from a lot to the nearest industrial zone (IN1 General Industrial, IN2 Light Industrial or IN3 Heavy Industrial) as indicated in the NSW Department of Planning and Environment (DPE) zoning layer.

Calculating the distance to a highway on/off ramp provides a good example of the use of the *MinDistance* formula with a *where* clause (equation 5.4). In the calculation of distance from each lot to the nearest highway on/off ramp, equation 5.4 queries a roads layer of the NSW Digital Topographical Database (DTDB) where on/off ramps were indicated as class subtype 2:

MinDistance ([Layer:RoadSegment],   [5.4]
    Where ([Attribute:RoadSegment:ClassSubtype] = 2))

It is important to note the *MinDistance* formula calculated the direct distance from the lot to a feature in another data layer. Experimentation with road network-based distance calculations indicated differences between direct and network distance measures were negligible. While computations based on road network distance are a theoretical improvement over direct distance calculations, the objective of minimising computation time and keeping calculations 'rapid' meant the latter were employed when performing the interactive, exploratory analysis.

If then else formulas were used to evaluate a condition and return a specified value if that condition is true. The independent variables 'unit', 'semi' (detached) and 'house', indicated property type and were calculated as dummy variables using a two-stage CommunityViz dynamic attribute calculation. The first stage used an *If then else* formula to aggregate the numerous property types present in the APM data. Property types indicated as houses and cottages were aggregated as 'house', units and studios as 'unit' and semis, duplexes, terrace houses villas and townhouses were grouped as 'semi'. The second stage used another *if then else* formula to assign a dummy value for the dynamic attribute house, unit and semi using a formula identical (e.g. for house) to equation 5.5, slightly modified for unit and semi.

IfThenElse (If ([Attribute:Residential Lots:Proptype2] = "House"),   [5.5]
  Then (1),
  Else (0))

The height of the building attribute, a number indicating the maximum allowable building height, was also drawn into the lots layer using an *if then else* formula. The data indicate the maximum height of a building that may be erected in an area as measured in metres (NSW Spatial Data Catalogue 2017).

A *get from closest* formula (equation 5.6) employees a spatial test to populate a dynamic attribute by drawing features from another layer based on attributes or an additional spatial test. The number of data layers were aggregated using a *get from closest* formula including dwelling count, person count, median income and median rent. This type of formula works well when referring to spatially aggregated data. Person count is the population data from the 2011 Census at the Statistical Area 1 (SA1) Census geography. Dwelling count is a tally of residences at the same spatial scale. Median income and median rent are both weekly values measured in dollars from the 2011 Census at the SA1 spatial geography.

GetFromClosest ([Attribute:SA1_2011_Data:Person_cou])   [5.6]

The lot size variable indicates the area in square metres of individual lots within the residential lots layer. Lot size values were calculated automatically via ArcGIS. A year variable was developed to investigate whether there were any broad differences between sales data recorded in the two halves of FY 2015; the second half of 2014 in comparison with the first half of 2015. The variable F2015 was developed with a manual selection based on the APM attribute 'EventDate' indicating the year of the data the sale was recorded. This distinguished sales events recorded from 1 July 2014 to 31 December 2014 from those recorded 1 January 2015 to 30 June 2015.

TABLE 5.1 Input data

Description	Units	Min	Max	Mean	Std dev	Source
Sales price	A$	125000	9000000	789123	398516	APM
Number of bedrooms	no.	0	10	2.38	1.5	APM
Number of bathrooms	no.	0	8	1.375	0.94	APM
Lot size	m²	1.63	115474	2682	6771	DCDB (der)
Max building height permitted	m	3.5	90	14.3	11.9	DPE
Dwelling type (is house?)	d	0	1	0.43	0.495	APM
Have walk-in wardrobe?	b	0	1	0.056	0.23	APM
Have rumpus room?	b	0	1	0.066	0.25	APM
Have internal laundry?	b	0	1	0.33	0.47	APM
Have balcony?	b	0	1	0.34	0.475	APM
Distance to nearest train station	m	18.6	6016	1267	1141	DTDB (der)
Distance to nearest commercial or mixed-use precinct (zoning)	m	0	1506	285	257	DPE (der)
Distance to nearest commercial precinct (zoning)	m	0	2697	479	431	DPE (der)
Distance to nearest industrial precinct (zoning)	m	0	5014	960	971	DPE (der)
Distance to nearest primary school	m	0	3117	600	432	DTDB (der)
Distance to nearest secondary school	m	17.5	3651	1006	631	DTDB (der)
Distance to nearest community pool	m	43	9545	2699	2150	DTDB (der)
Distance to nearest highway on/off ramp	m	3.6	3908	1407	830	DTDB (der)
SA1 population count	no.	0	2647	531	380	Census2011
SA1 dwelling count	no.	0	1301	218	193	Census2011
SA1 median weekly household income	A$	0	2939	1386	428	Census2011
SA1 median weekly rent	A$	0	890	369	113	Census2011
Is in fiscal year 2015?	d	0	1	4140	0.451	APM

m=metres; d=dummy variable; b=Boolean variable; der=derived with CommunityViz from these datasets.

A summary of the data input into the hedonic price model along with descriptive statistics and the source of the data is presented in Table 5.1.

Hedonic price analysis

The second stage of the analysis was to develop a robust hedonic price model from the data. The econometric model was calculated using ordinary least squares (OLS) using GeoDa™ 1.8.16.4 (released 1 March 2017). The modelling, while preliminary, addressed a number of econometric problems frequently encountered in hedonic price modelling. Multicollinearity

TABLE 5.2 Hedonic regression results

Dependent variable = sales price

Independent variable	Coefficient (std. error)
Number of bedrooms	9,023* (3,914)
Number of bathrooms	76,935*** (5,769)
Lot size	2.3*** (0.47)
Max building height permitted	653* (286)
Dwelling type (is house?)	378,663*** (8,404)
Have walk-in wardrobe?	67,435*** (13,220)
Have rumpus room?	30,361* (12,659)
Have internal laundry?	-46,013*** (6,882)
Have balcony?	-51,544*** (7,274)
Distance to nearest train station	-51.72*** (3.27)
Distance to nearest commercial or mixed-use precinct (zoning)	-48.67** (17.69)
Distance to nearest commercial precinct (zoning)	29.85** (9.51)
Distance to nearest industrial precinct (zoning)	39.24*** (4.92)
Distance to nearest primary school	40.63* (9.47)
Distance to nearest secondary school	24.85*** (6.04)
Distance to nearest community pool	39.75*** (2.97)
Distance to nearest highway on/off ramp	-23.79*** (4.17)
SA1 population count	-234.7***(52.1)
SA1 dwelling count	455.3*** (106.4)
SA1 median weekly household income	109.1*** (11.14)
SA1 median weekly rent	201.3*** (38.2)
Is in fiscal year 2015?	81,790*** (5,831)
Constant	138,259*** (15,702)
Adj R²	0.518
Degrees of freedom 9147	

* p < 0:05, ** p < 0:01, *** p < 0:001.

among nominally independent variables is a common and well-known problem in hedonic regressions (de Hann and Diewert 2013), as many property characteristics move in concert with higher or lower sales prices. Prior to selecting variables for analysis in the hedonic regression, we calculated the influence of each potential independent variable on multicollinearity using variance inflation factors (VIF). We then omitted variables with VIF values greater than or equal to 10. A hedonic price model was then developed from the remaining sales price and property, neighbourhood and proximity characteristics aggregated within the lots layer. OLS results are presented in Table 5.2.

Overall the fit of this model is sufficiently robust for this analysis with an adjusted r² of just over 0.5. In addition and equally important for the purpose of this modelling, the majority of the independent variables are statistically significant and have the expected sign. The number of bedrooms, number of bathrooms, lot area, house (as opposed to unit or semi-detached) and height limit are all positively signed, indicating a direct relationship between these characteristics and higher sale prices. Other property characteristics directly associated with higher sales prices include the presence of a walk-in wardrobe or rumpus room. Neighbourhood characteristics including dwelling count, median income and median rent are also positively associated with price. In contrast, the inclusion of an internal laundry and/or a balcony had a negative influence on price; we assume because these property characteristics are most often associated with apartments or units which are usually of lower values than detached houses, primarily due to the absence of a dedicated land component.

The distance to train station, distance to highway on/off ramp and distance to commercial or mixed-use zones are signed negatively. As these are distance measures, the negative sign indicates proximity commands a higher price, increasing the distance from the feature to the lot results in a lower price. In contrast, distance to the industrial zones is a positively signed distance-based characteristic. This indicates housing prices increase with further distance from the industrial zones. Importantly, the modelling indicates that within the study area there is a willingness to pay for proximity to a train station.

The coefficients in Table 5.2 can be used to estimate property values for all properties where the independent variables are known. As outlined below, when the attribute values are adjusted, then the estimated property value will also be adjusted in line with the modelling. This allows the uplift in value caused by the addition of a nearer train station, for example, to be estimated.

Completing the decision-making framework

The decision framework of the prototype RAISE tool was based on the spatial aggregation of input layers within the lots layer, and quantified statistically significant relationships between different layers and residential sales prices in the study area determined with a hedonic regression model. After running the regression, remaining tasks prior to completing the decision framework included developing a dynamic attribute for hedonic price, developing a set of scenarios to facilitate interactive exploratory analysis of transportation infrastructure development and developing a set of indicators for evaluating those scenarios. The dynamic attribute for hedonic price uses the geographic data aggregated in the lots layer and the coefficients presented in Table 5.2 to calculate a modelled price for each residence within the study area. This is an example of a coefficient formula where regression coefficients are multiplied by data attributes to model hedonic prices.

The set of scenarios designed to facilitate interactive exploratory analysis of transportation infrastructure is initially limited to two scenarios. There were a current conditions scenario and a future transport corridor scenario, where users of the RAISE toolkit can 'drag and drop' new train stations and the value uplift for all residential properties is calculated on the fly as driven by the coefficients calculated in the hedonic price modelling for the study area.

The impacts of the different scenarios are evaluated quantitatively using indicators. Indicators are formula-based map and data layer summary values (Walker and Daniels 2011; Lieske *et al.* 2015) used to highlight similarities and differences in scenarios; in this case transportation infrastructure development and value uplift. The formula-based indicators developed for this analysis include counts of residential lots within specified distance bands from train stations (i.e. 400m, 800m and beyond), sales prices (i.e. existing data) and modelled hedonic prices within these specified distances bands. Indicators were also calculated for average value uplift and total value uplift within the indicated distance bands.

Project implementation: interactive scenario exploration with the toolkit

Once a decision framework was constructed, the process of sketching scenarios, exploring alternatives and evaluation impacts is referred to as an 'analytical process or geodesign' (Walker and Daniels 2011, p.28). It is at this stage that citizens, planners, elected officials, decision-makers and other parties can explore design alternatives with immediate feedback on the consequences of their ideas. The scenarios and selected indicators for the prototype RAISE toolkit are presented in Figure 5.3. Interactivity was facilitated with the CommunityViz sketch tools which allow easy editing of GIS layers to add, subtract and move transportation infra-structure such as train stations or highway on/off ramps. Figure 5.3 uses the CommunityViz 'compare scenarios' view to present a base scenario representing current conditions and a hypothetical extended north metro line scenario designed for interactive experimentation for new train station alternatives. As shown in Figure 5.3, the hypothetical extended north metro line scenario explores the development of a new train line and five new train stations (encircled in Figure 5.3) in the north-western portion of the study area.

Scenarios in CommunityViz are spatially determined impact models. When working with the CommunityViz sketch tools feature changes such as additional train stations or additional residences will result in automatic recalculation of all relevant dynamic attributes and indicators, much the same way as changing a value in a spreadsheet will result in recalculation of other linked values. For example, the lot count indicator in Figure 5.3 shows a base scenario with 1,921 residential lots located within 400m of a train station, 2,246 lots located within 800m of a train station and 5,004 lots located in the remainder of the study area. By including a number of additional train stations, the hypothetical extended north metro line scenario CommunityViz calculates an increase in the number of residential lots within 400m of a train station to 2,125 lots, an increase in the number of lots within 800m of a train station to 2,748 and, because no new residences are included, a reduction in the remainder of the study area to 4,298 lots.

Hedonic regression-based price model results for the two scenarios are also presented in Figure 5.3. The base scenario shows the average modelled price of lots located within 400m of a train station is A\$624,864, lots located between 400m and 800m are A\$700,078 and lots located beyond 800m of a train station priced at A\$820,517. The hypothetical extended north metro line scenario hedonic results indicates lots located within 400m of station are priced at A\$640,913, lots located between 400m and 800m are priced at A\$713,688 and lots located beyond 800m of a train station priced at A\$861,359.

Value uplift to the addition of the train stations can be calculated by subtracting the hedonic model price of the base scenario from the hypothetical extended north metro line scenario for lots impacted by the new train stations. As shown in Figure 5.4, the average value uplift for lots located in the 400m distance band is A\$16,049; A\$13,611 for lots located in the 800m dis-tance band and A\$40,842 for lots located beyond 800m from the new train stations. The total value uplift is also presented in Figure 5.4, where total value uplift within 400m of the new train stations is modelled at A\$3,274,026. The value uplift between 400m and 800m of the new train stations is A\$6,832,563. The value uplift beyond 800m from the new stations, where property values were raised because of the new stations, equating to 649 properties, totalled A\$26,506,459. The total value of value uplift for residential land resulting from the five new stations is modelled to be A\$36,613,048.

The research question explored here is: *Can a rapid spatial analysis decision-support platform be created to support land-use planners' exploration of value uplift scenarios in real-time?* By demon-strating the use of GIS and PSS to aggregate data within a single layer using dynamic attributes,

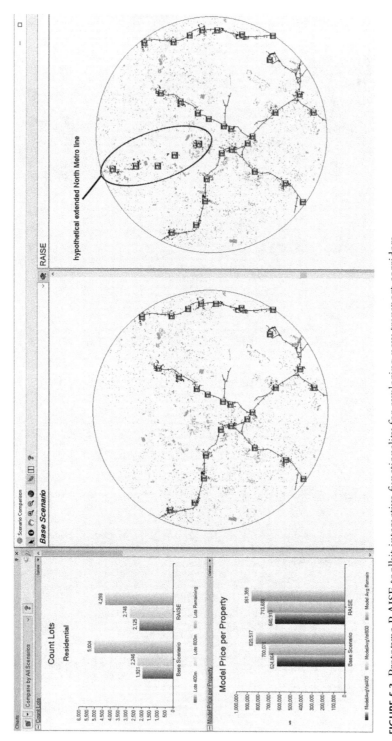

FIGURE 5.3 Prototype RAISE toolkit interactive functionality for exploring new transport corridors

Source: authors.

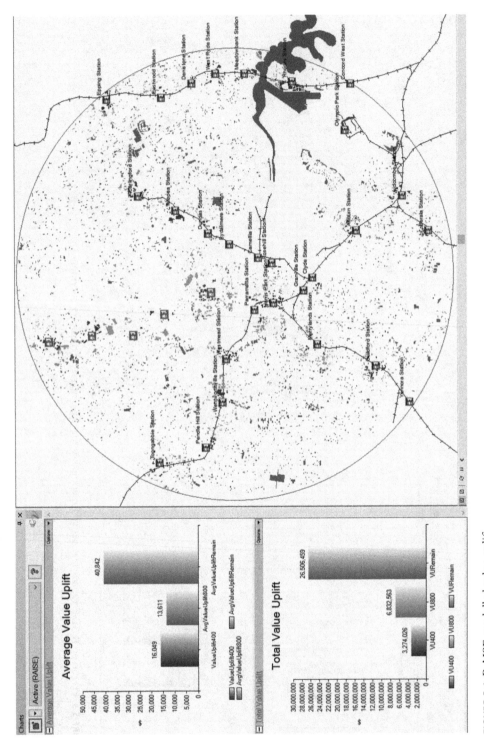

FIGURE 5.4 RAISE modelled value uplift
Source: authors.

developing a hedonic price model based on those data, bringing the coefficients from the econometric modelling results into an interactive spatial scenario planning environment, enabling sketching of development alternatives and quantifying the impact of those alternatives, then we are able to create a tool to support exploration of value uplift scenarios in real-time.

Conclusion

In order to make sense of a diverse and rapidly changing data environment, tomorrow's spatial data tools must help decision-makers not simply to combine and analyse data, but to explore data in new collaborative ways that harness collective intelligence and foster transparency in decision-making. The contribution of the project described in this chapter is a set of methods for the rapid quantification of value uplift based on integration of hedonic price modelling into a GIS and PSS platform. The approach to quantifying value uplift present here is to specify a hedonic price model using geographic information systems (GIS) and planning support systems (PSS) attributes. GIS and PSS are used to develop a data set of property, neighbourhood and environmental characteristics relevant to hedonic price modelling in the study area. An econometric model has been used to determine statistically significant determinants of price. The marginal influence of significant property characteristics (i.e. coefficients) are then incorporated within an interactive scenario explorative process which enables sketching new transportation infrastructure with real-time feedback on changes in value uplift.

From this research we are able to draw a number of conclusions relevant to the RAISE approach and the development of underlying hedonic pricing models in the context of the Greater Sydney Area. This analysis demonstrated the willingness to pay for access to transit, specifically for access to commuter rail stations in the Parramatta City Council Area of Western Sydney. Model results indicated an indicative total value uplift of A\$3,274,026 within 400m of the new train stations, value uplift of A\$6,832,563 between 400m and 800m of the new train stations and value uplift of A\$26,506,459 beyond 800m from the new stations. The grand total of value uplift from the five new stations in the hypothetical extended north metro line is modelled to be A\$36,613,048.

We also conclude that data available in the Greater Sydney Area on the attributes of houses and land, location characteristics and neighbourhood characteristics are suitable for development into a hedonic price model. This includes the caveat noted by Palmquist (1991) that data on structural characteristics are typically more reliable than data on neighbourhood and environmental variables. Future hedonic price models developed for Sydney may include building characteristics that are not yet widely available for the study area. These characteristics include building footprints, number of floors, building age, building materials and other aspects of construction quality.

There are also a number of next steps and promising directions for future research. Scenario-based PSS offer a number of related opportunities for exploring transportation infrastructure and other development not explored in detail in this chapter. The range of scenarios that could be explored may be extended beyond the current conditions scenario and a hypothetical extended north metro line presented here. Scenario planning allows different configurations of infrastructure and changes to residential stock to be built and compared. Decision-makers and planners can potentially explore any number of different locations for train lines and stations as well as build scenarios for other residential, commercial and industrial land uses, allowing quantitative and visual assessment of impacts. This allows detailed exploration of planning alternatives prior to construction that facilitates both optimisation and public-engagement in the planning process. The power of a rapid analytics approach is to enable the interactive exploration of a number of *what if?* scenarios to get a comparative analysis of the likely financial implications of infrastructure decisions.

This study focused on extending commuter rail infrastructure by looking at development of new train stations in Western Sydney. The economic results also allow for the exploration of alternatives for new highway on/off ramps. The decision framework is also built to allow evaluation of new and different types of residential development. This is typically done with typologies for single family homes, multi-family homes, low-rise apartments and high-rise apartments that associated plausible attributes with building types. Similar to the number of scenarios that may be considered, there is no limit to the number of building typologies that could be incorporated.

In the context of Sydney there are a number of infrastructure projects that could benefit from consideration of associated value uplift and potential value capture opportunities. These include the Sydney CBD to Parramatta metro rail line (O'Sullivan 2017), fast metro rail to the new Sydney airport (Paterson 2017) as well as intercity high speed rail (Carey 2017). In New South Wales, probable construction of a second airport in Western Sydney has resulted in a recognised need to invest in transportation infrastructure. The NSW Government has committed to over A\$73 billion worth of projects in the four years preceding and through 2019–20 (NSW Department of Industry 2017). Using a value uplift method on just these NSW projects alone could provide a strong evidence base to support value capture policies which could in turn save money for the NSW taxpayer.

The econometric modelling presented here was simplified in order to as clearly as possible demonstrate the links between geographic data, hedonic price model components and scenario planning. Any decisions on value uplift on value capture should incorporate modelling that fully addresses econometric issues inherent in hedonic price modelling, including serial correlation, heteroscedasticity and spatial autocorrelation. The econometric model may also be refined by exploring non-linear relationships in distance-based property characteristics. These surfaces are easy to develop, may be included in OLS and more sophisticated regression models and also may easily be incorporated in a 'rapid' price evaluation setting. Future models may also explore broader time-scales and trends in the real estate market.

As cities continue to grow there is a need for smarter data-driven tools which can model future scenarios and identify where value creation is likely to occur. The RAISE toolkit presented in this chapter provides one such PSS which can assist planners and decision-makers in the early identification of value uplift as driven by new transport infrastructure and provide them the evidence required to develop value capture policies to support smarter city planning.

Acknowledgements

The support of the CRC-SI is acknowledged for funding this research project under Program 2 Rapid Analytics.

References

Australian Bureau of Statistics (2017). Regional population growth, Australia, 2015–16. Retrieved from: www.abs.gov.au/ausstats/abs@.nsf/mf/3218.0. Accessed 6 April 2017.

Australian Government Department of the Prime Minister and Cabinet (2016). Smart Cities Plan. Retrieved from: https://cities.dpmc.gov.au/smart-cities-plan. Accessed 23 August 2017.

Bartholomew, K. and Ewing, R. (2011). Hedonic price effects of pedestrian- and transit-oriented development. *Journal of Planning Literature*, 26, 18–34; doi:10.1177/0885412210386540.

Bastian, C.T., McLeod, D.M., Germino, M.J., Reiners, W.A. and Blasko, B.J. (2002). Environmental amenities and agricultural land values: a hedonic model using geographic information systems data. *Ecological Economics*, 40, 337–349; doi:10.1016/S0921-8009(01)00278-6.

Carey, A. (2017). Act now on high-speed rail or pay heavy price later: Infrastructure Australia. *Sydney Morning Herald*. Retrieved from: www.smh.com.au/national/act-now-on-highspeed-rail-or-pay-heavy-price-later-infrastructure-australia-20170706-gx63fz.html Accessed 23 August 2017.

Cervero, R. and Murakami, J. (2009). Rail and property development in Hong Kong: experiences and extensions. *Urban Studies*, 46, 2019–2043; doi:10.1177/0042098009339431.

Chhetri, P., Han, J.H. and Corcoran, J. (2009). Modelling spatial fragmentation of the Brisbane housing market. *Urban Policy and Research*, 27, 73–89.

Costello, G. and Watkins, C. (2002). Towards a system of local house price indices. *Housing Studies*, 17, 857–873; doi:10.1080/02673030216001.

de Haan, J. and Diewert, E. (2013). Hedonic regression methods. *Handbook on Residential Property Price Indices*. Eurostat, 49–64; doi:10.1787/9789264197183-7-en.

Davison, G., Han, H. and Liu, E.J. (2016). The impacts of affordable housing development on host neighbourhoods: two Australian case studies. *Housing and the Built Environment*; doi.org/10.1007/s10901-016-9538-x.

Elaurant, S and Louise, J. (2015). Politics, finance and transport – megaprojects in Australia [online]. *Proceedings of the Royal Society of Queensland*, 120, 31–45.

Geertman, S., Allan, A., Pettit, C., Stillwell, J. (2017). Introduction to 'Planning Support Science for Smarter Urban Futures'. In *Planning Support Science for Smarter Urban Futures*, 1–19. Springer International Publishing.

Han, J. H., Kim, J.Y. and Kim, J. (2017). Dynamics of Housing Mobility in Australian Metropolitan Areas, 2001–2010: A Longitudinal Study. *Urban Policy and Research*, 35, 122–136.

Higgins, C. and Kanaroglou, P. (2017). Rapid transit, transit-oriented development, and the contextual sensitivity of land value uplift in Toronto. Urban Studies; doi:10.1177/0042098017712680.

Infrastructure Australia (2016). Capturing value: advice on making value capture work in Australia. Infrastructure Australia. Retrieved from: http://infrastructureaustralia.gov.au/policypublications/publications/files/Capturing_Value_Advice_on_making_value_capture_work_in_Australia-acc.pdf. Accessed 15 May 2017.

Lieske, S.N. and Gribb, W.J. (2012). Modeling high-resolution spatiotemporal land-use data. *Applied Geography*, 35, 283–291; doi:10.1016/j.apgeog.2012.06.001.

Lieske, S.N. and Hamerlinck, J.D. (2015). Integrating planning support systems and multicriteria evaluation for energy facility site suitability evaluation. *Journal of the Urban and Regional Information Systems Association*, 26, 13–24.

Lieske, S.N., McLeod, D.M. and Coupal, R.H. (2015). Infrastructure development, residential growth and impacts on public service expenditure. Appl. *Spatial Analysis*, 8, 113–130; doi:10.1007/s12061-015-9140-8.

McGuirk, P. and Argent, N. (2011). Population growth and change: implications for Australia's cities and regions. *Geographical Research*, 49, 317–335; doi:10.1111/j.1745-5871.2011.00695.x.

McIntosh, J., Trubka, R. and Hendricks, B. (2016). Transit and urban renewal value creation. Tech. rept. Luti Consulting. Retrieved from: www.luticonsulting.com.au/wp-content/uploads/2013/12/Sydney-Transit-and-Urban-Renewal-Value-Creation-Report.pdf.

Minister for Urban Infrastructure (2017). Government commits up to $5.3 billion to build Western Sydney Airport. Retrieved from: http://minister.infrastructure.gov.au/pf/releases/2017/may/pf015_2017.aspx. Accessed 24 May 2017.

Mulley, C. (2014). Accessibility and residential land value uplift: identifying spatial variations in the accessibility impacts of a bus transitway. *Urban Studies*, 51, 1707–1724; doi:10.1177/0042098013499082.

Murray, C.K. (2016). Land value uplift from light rail. Retrieved from: https://papers.ssrn.com/sol3/papers.cfm?abstract_id=2834855.

NSW Department of Industry (2017). Infrastructure and construction. Retrieved from: www.industry. nsw.gov.au/invest-in-nsw/industry-opportunities/infrastructure-and-construction. Accessed 29 August 2017.

NSW Spatial Data Catalogue (2017). Standard Instrument Local Environmental Plan (LEP) – Height of Building (HOB). Retrieved from: https://sdi.nsw.gov.au/sdi.nsw.gov.au/catalog/search/resource/details.page?uuid=%7B83B5C6B8-3BE0-4014-8CB1-37734462510D%7D. Accessed 28 March 2017.

NSW (2017). Glossary. New South Wales Government Land and Property Information. Accessed 19 May 2017.

O'Neill, P.M. (2010). Infrastructure Financing and operation in the contemporary city. *Geographical Research*, 48, 3–12; doi:10.1111/j.1745-5871.2009.00606.x.

O'Sullivan, M. (2017). Cost of new metro line from Sydney CBD to Parramatta set to top $12.5 billion. *Sydney Morning Herald*. Retrieved from: www.smh.com.au/nsw/cost-of-new-metro-line-from-sydney-cbd-to-parramatta-set-to-top-125-billion-20170627-gwzd5d.html. Accessed 23 August 2017.

Páez, A., Scott, D.M. and Morency, C. (2012). Measuring accessibility: positive and normative implementations of various accessibility indicators. *Journal of Transport Geography*, 25, 141–153; doi:10.1016/j.jtrangeo.2012.03.016.

Palmquist, R.B. (1991). Hedonic methods. In: Braden, J., Kolstad, C. (eds.), *Measuring the Demand for Environmental Quality*, 77–120. Elsevier Science, Amsterdam.

Paterson, I. (2017). New Sydney Metro service will link Western Sydney Airport to the rest of the city. *Daily Telegraph*. Retrieved from: www.dailytelegraph.com.au/projectsydney/new-sydney-metro-service-will-link-western-sydney-airport-to-the-rest-of-the-city/news-story/e5baf9f72889d69f350 2b2818d97e52a. Accessed 29 August 2017.

Pettit, C.J., Klosterman, R.E., Delaney, P., Whitehead, A.L., Kujala, H., Bromage, A. and Nino-Ruiz, M. (2015). The online what if? planning support system: a land suitability application in Western Australia, *Applied Spatial Analysis and Policy*, 8(2), 93–112.

Randolph, B. and Tice, A. (2014). Suburbanizing disadvantage in Australian cities: sociospatial change in an era of neoliberalism. *Journal of Urban Affairs*, 36, 384–399; doi:10.1111/juaf.12108.

Randolph, B. and Tice, A. (2017). Relocating disadvantage in five Australian cities: socio-spatial polarisation under neo-liberalism. *Urban Policy and Research*, 35, 103–121; doi:10.1080/08111146.2016.1221337.

Rosen, S. (1974). Hedonic prices and implicit markets: product differentiation in pure competition. *Journal of Political Economy*, 82, 34–55.

Smith, J.J. and Gihring, T.A. (2017). *Financing Transit Systems Through Value Capture: An Annotated Bibliography*. Victoria Policy Institute. Retrieved from: www.vtpi.org/smith.pdf. Accessed 12 April 2017.

Terrill, M. and Emslie, O. (2017). *What Price Value Capture?* (No. 0987612123). Grattan Institute.

Thie, J. (2008). Cadastre. In K. Kemp (ed.), *Encyclopedia of Geographic Information Science*, 15–19. Los Angeles: Sage.

Turnbull, M. (2016). Speech to the National Cities Summit. Melbourne. Prime Minister of Australia. Retrieved from: www.pm.gov.au/media/2016-04-29/speech-national-cities-summit-melbourne-0. Accessed 23 August 2017.

Van Kooten, G.Cornelis. (1993). *Land Resource Economics and Sustainable Development*. University of British Columbia Press: Vancouver.

Waddell, P. (2002). UrbanSim: Modeling urban development for land use, transportation, and environmental planning. *Journal of the American Planning Association*, 68(3), 297–314; doi: 10.1080/01944360208976274.

Wadley, D., Elliott, P. and Han, J.H. (2017). Modelling homeowners' reactions to the placement of high voltage overhead transmission lines. *International Planning Studies*, 22(2), 114–127.

Walker, D. and Daniels, T.L. (2011). *The Planners Guide to CommunityViz: The Essential Tool for a New Generation of Planning*. Planners Press, American Planning Association, Chicago.

6

COMMERCIAL OFFICE PROPERTY AND SPATIAL ANALYSIS

Jian Liang and Richard Reed

Introduction

This chapter investigates the importance of location and mapping with reference to office buildings located in a central business district (CBD). More specifically, it examines the contribution of spatial characteristics when undertaking the challenging task of constructing a price index for the transactions of whole office buildings, which for the balance of this chapter will simply be referred to as office buildings. To date, there have been limited attempts to construct an accurate property index which is arguably essential for stakeholders and the market-place to understand movements in both the property market and also in the broader economy. This chapter also includes a case study based on all transactions of office property buildings in the Melbourne CBD between 2000 and 2015. The starting point for this research is the expectation that there would be significant spatial dependency in the transactional market of office buildings.

Arguably the benefit of incorporating the spatial effect by using a spatial lagged and spatial error model will improve the accuracy of an office building price index. Furthermore, the findings should highlight the importance of the issue of spatial autocorrelation in the estimation of valuation models and price indexes for office buildings. The findings will also assist stakeholders, including valuers, investors and market regulators, to improve their understanding of movements in the office property buildings transactional market via an increased understanding of the importance of spatial characteristics. Accordingly, the research question for this chapter is as follows: *To what extent does spatial dependency exist in the transactional market for office buildings?*

This research develops a transactional price index for the office market and tests how incorporating spatial dependency issues into the estimation of the index can improve the level of accuracy. The research is limited to the office building market and is ideally applied to a central business district or an agglomeration of office buildings; however, with allowances the methodology can be applied to related land uses and similar locations.

Office building indexes

Since the global financial crisis in 2007 there has been an upturn in the level of interest in office properties (Lizieri and Pain 2014). This has placed pressure on researchers to develop new methods of analysing office property markets to identify and better understand the individual supply and demand drivers. In the property and real estate market-place, one of the most under-researched areas has traditionally been the locational attributes of the property, which from a property perspective is commonly referred to as 'mapping' or 'spatial' characteristics. Spatial dependency is when everything in space is related, but the relatedness of things generally decreases with distance (Tobler 1970). This relationship is also applicable to property, as there is a strong relationship between the role of spatial dependency and real estate asset value (Andersson and Gråsjö 2009). For the transactional market of office units it has been shown that controlling for spatial dependency improves the accuracy of valuation models and corresponding price indexes (Tu *et al.* 2004; Maury 2009). However there are substantial differences between constructing an index for (a) smaller areas in an office building being smaller office units (e.g. part of one floor, an entire floor) in comparison to (b) a whole office building; note that the market for (b) is already well established while the market for (a) has received very little attention in the literature.

Practitioners may rely on comparable office units located close by when valuing individual office units; however they do not rely on nearby comparable transactions of entire office buildings when valuing entire office buildings, partly due to the absence of sales of entire buildings and also their heterogeneity. Furthermore, factors such as the land size and land value which influence the entire office building prices are usually not usually factored into the valuation process for individual office units. There is no evidence to date supporting the existence of spatial dependency in the transactional market of whole office buildings, even if the existence of spatial dependency in the transactional market of office property units has been already been confirmed (Tu *et al.* 2004).

This study is based on the hedonic model approach, which is widely used in constructing property indexes since it can effectively control for the heterogeneity of property characteristics by incorporating all property-related attributes into an estimation (Ericson *et al.* 2013). In addition, the hedonic model can produce a better price index for office property than the repeat sales method (Shiller 1991) because the amount and frequency of transactions of office property are usually much smaller than for residential property. In other words, the number of repeat sales of traded office property is too low to facilitate empirical research. Also, the use of a repeat sales method would increase the problem of sample selection bias (Shiller 1991). Traditionally entire office properties (i.e. land and buildings) do not usually have their appraisal value provided by the government, for example, in the same manner as residential property in many locations; therefore the sale price appraisal method cannot be readily applied (Shi, Young and Hargreaves 2009). It should also be noted that the office property market usually has a more standardised structure compared to other types of property, such as residential, industry and retail property. In this context, the unified structure of office property requires fewer controlling variables in the hedonic model and can help to mitigate the possible bias caused by omitting variables. On this basis the hedonic model was selected to construct the price index for office properties to examine spatial dependency.

Spatial dependency of office property

Earlier studies conducted empirical tests to examine how the transactional price of office property is influenced by various factors, which can be broken down into three types:

(a) Physical characteristics and related factors;
(b) Locational factors and other factors such as the performance of the property itself (Henneberry and Mouzakis 2014); and
(c) Regulatory factors (Smith *et al.* 2000).

Previous research has confirmed the relevant influential factors related to physical characteristics of office buildings, including:

* Quality/grading of the building (Dermisi and McDonald 2010; Eichholtz *et al.* 2010);
* Size of traded office property (Nappi *et al.* 2007);
* Age of the office property (Dermisi and McDonald 2010); and
* Design/energy rating (Fuerst and McAllister 2011; Bonde and Song 2013; Eichholtz *et al.* 2010).

Note that this research will include these factors into the estimation of regression model and index, except for the design/energy rating which is highly correlated with the grading and the age of the office property and will cause problems associated with multicollinearity. In addition to physical characteristics it has been shown that transfer price of office property is impacted by accessibility to public transport (Kim *et al.* 2015; Nitsch 2006) and varies between different submarkets (Dunse *et al.* 2001). Building upon the established framework, this research will incorporate the distance to key train stations as a controlling variable in the analysis. Arguably, including the distance to key locations can capture some of the geographical features but it cannot effectively control for the issue of spatial dependency (Andersson and Gråsjö 2009). With reference to the valuation discipline, the issue of spatial dependency has been investigated for different type of properties including residential (Dubin 1992), retail (Lee and Pace 2005; Liang and Wilhelmsson 2011) and rural (Huang *et al.* 2006). With reference to office property valuation, Maury (2009) examined office property unit transactional data in Paris and incorporated spatial dependency in the model. Tu *et al.* (2004) utilised a database containing office unit transactional data in Singapore to test for the impact of spatial dependency on the valuation of office property, where the findings indicated that controlling the level of spatial dependency can improve the accuracy of valuation of the office property units and the estimation of a corresponding office unit index.

The transactional market for office buildings differs from the transactional market for individual office units, as the office property units in the same office building are usually transacted at a similar price since practitioners rely on the comparable transactions. Thus, the existence of spatial dependency in the transactional market of property units is easy to understand, given that the practitioners reply on transactional price of comparable office units located in the same office building to evaluate an individual office unit. However, the turnover rate of office buildings in the market-place is lower than that for office property units, so practitioners usually have difficulty in identifying comparable transactions nearby to provide a reference point for the transaction price. Furthermore, the process of valuing whole office buildings differs substantially from the approach for the valuation of individual office property units; for example, the value of the whole office property building cannot be calculated by proportionating the transactional price of office units in the building by the total building area. Arguably, if this occurred, then many other important factors such as the land size/value and variations between office units in the same building are ignored. Therefore, the existence of spatial dependency in the transactional market of whole office property buildings is uncertain, even if the existence of spatial dependency in the transactional market of office property units is confirmed. In contrast to examining the transactions of office units, it is relatively difficult to find comparable transaction of similar office buildings located nearby. In other words, practitioners need to mainly rely on other methods, predominantly the income approach, to determine the value of the whole office building. The value of investment grade property, such as office property, must be determined by its income-producing capacity, which is directly related to rental income (Reed 2015). Furthermore, it is accepted that the rental income of office property is related to location and notably to spatial patterns, confirming the existence of spatial dependency for office rent distribution (Jones and Orr 2004; Bollinger *et al.* 1998; Sivitanidou 1995). It is argued that both the income-producing capacity and also the value/transactional price of an entire office building should follow a certain spatial pattern in a city where spatial dependency exists.

Research question

For this chapter the research question is: *To what extent does spatial dependency exist in the transactional market for whole office buildings*

The main contribution to the existing literature relates to the relationship between (a) the variable of spatial dependency and (b) transactions of whole office buildings. Due to the nature of the office market and the higher value of whole office buildings (in comparison to smaller areas), most of the attention in the office market is placed on the transfer of tenure associated with smaller net lettable areas in office buildings, e.g. part of a floor or an entire floor in an office building. Previously the main barrier has been the availability of data and accessing an adequate quantity of sales, but now both challenges have been overcome with the methodology adopted for this research.

Research methods

This study uses a hedonic model to construct a property price index for office buildings. With this approach, it is accepted that a hedonic model considers each whole office building as a normal commodity for which the market value is determined by attributes such as quality, area, age and location (Ericson *et al.* 2013). In addition, the marginal effect of each attribute on the price of the office property is estimated as coefficients for each independent variable in the hedonic model. Accordingly the baseline hedonic model is expressed as follows:

$$P_{i,t} = \beta_0 + CX_{i,t}\beta_1 + LX_{i,t}\beta_2 + TD_t\beta_3 + \varepsilon_{i,t} \qquad [6.1]$$

The dependent variable P is the transactional price per m^2. The vector of coefficients X in model 6.1 is decomposed into three types of independent variables in model 6.2: CX are associated with the physical attributes of the property; LX indicate the locational attributes of the property; *TD* is the dummy variable indicating in which year the property is sold. The index number is the estimated coefficient β_3 which measures how the property's transaction price changed between different years (Ericson *et al.* 2013). The variables for physical attributes (*CX*) in the model include the building quality grading (i.e. Premium, A, B, C, D, where Premium is the highest quality) of the office space, the age of the building, the area of the building and the total land area. The variables for locational attributes (*LX*) include distance to the office property per square m^2 and the distance to train stations in the CBD. We use 16 dummy variables for the years from 2000 to 2015 in which the transactions were recorded, and then build the index based on these 16 dummy variables.

Spatial dependency

The spatial dependency can be measured by using a spatial weights matrix to describe the relative geographical location of each observation (Andersson and Gråsjö 2009; Maury 2009; Tu *et al.* 2004; Tobler 1970). The spatial weights for matrix W is expressed as follows:

$$W = \left(\begin{bmatrix} w_{1,1} & w_{1,2} & \cdots & w_{1,j-1} & w_{1,j} \\ w_{2,1} & w_{2,2} & & w_{2,j-1} & w_{2,j} \\ \vdots & & \ddots & & \vdots \\ w_{i-1,1} & w_{i-1,2} & \cdots & w_{i-1,j-1} & w_{i-1,j} \\ w_{i,1} & w_{i,2} & & w_{i,j-1} & w_{i,j} \end{bmatrix}\right) \qquad [6.2]$$

In this matrix the total number of observations is equal to the total number of rows is *i* and the total number of columns is *j*. The element of the matrix "w" indicate the relative geographic location of each observation by calculating the inverse proportion of the geographic distance between observations. For example, if the geographic distance between observation "2" and observation "5" is 10, the value of elements "$w_{2,5}$>" and "$w_{5,2}$>" in this matrix (6.3) will be 0.1 (1/10). We further standardise each row by dividing each element of each row by the total of the row, then the spatial weights matrix "W_{Sd}>", which measures the spatial dependency (i.e. proximity of observations to each other).

$$w_{Sd} = \begin{bmatrix} \dfrac{w_{1,1}}{\sum_j^1 w_{1,j}} & \cdots & \dfrac{w_{1,j}}{\sum_j^1 w_{1,j}} \\ \vdots & \ddots & \vdots \\ \dfrac{w_{i,1}}{\sum_j^1 w_{i,j}} & \cdots & \dfrac{w_{i,j}}{\sum_j^1 w_{i,j}} \end{bmatrix} \qquad [6.3]$$

We can also set the limit of distance for spatial dependency in the matrix. For example, we can assume that the observations with geographic distance larger than 10 do not constitute spatial dependency to each other.

$$w_{i,j} = \begin{cases} \dfrac{1}{d_{i,j}}, & \text{if } d_{i,j} \leq 10 \\ 0, & \text{otherwise} \end{cases} \qquad [6.4]$$

In equation 6.4 reference to "$d_{i,j}$" is the distance between observations i and j. We further build the spatial lagged model based on this spatial weights matrix if the spatial dependency exists for the dependent variable "P" in model 6.1. Therefore, the transactional price of the office property is more comparable to the price of office property buildings located in closer proximity than office property located further away. In this case the transactional price can be explained by the price of nearby property where the baseline model [6.1] then becomes:

$$P_{i,t} = \beta_0 + \rho W P_{it} + X_{i,t}\beta_1 + \varepsilon_{i,t} \qquad [6.5]$$

Therefore "W" is the weighted matrix and "ρ" is the coefficient of the spatial dependency i.e. the weighted average price of office property located in close proximity. If there are other factors which are not included in the model but also causing spatial dependency, we estimate the spatial dependency of error term by using the spatial error model. In equation 6.6 the "$\varepsilon_{i,t}$" is the error term containing spatial dependency and the "μ" is the error term without spatial dependency. We can also calculate the Moran's I statistics and Lagrange multiplier statistics to test for the existence of spatial dependency.

$$P_{i,t} = \beta_0 + X_{i,t}\beta_1 + \varepsilon_{i,t}, \text{ where } \varepsilon_{i,t} = \lambda W \varepsilon_{i,t} + \mu \qquad [6.6]$$

Analysis and case study – Melbourne CBD

This research examines real estate and spatial data relating to all office building transactions observed in the Melbourne central business district (CBD) between 2000 and 2015; the total number of observations during this time period was 289. Table 6.1 summarises the statistical characteristics of the database.

As shown in Table 6.1, the average age of the office buildings was 55 years. The average transactional price was US$48 million and the average building area and average land were 12,667 and 1,775 m^2 respectively. The B grade office properties represented the biggest proportion (47%) of buildings, followed by C grade (25%) and then A grade office (16%) properties. The transactions were distributed relatively evenly across the time-frame; for example, the proportion in 2001 represented 10% of the total database and in 2015 it represented 9%. Note that the statistical characteristics of the data were consistent with previous research (see Maury 2009; Tu et al. 2004). The correlation coefficients between variables which will be used in the model are presented in Table 6.2. It is evident that all observations were not strongly correlated with each other; the individual variance inflation factors (VIF) statistics for all these variables and the average VIF results were not larger than 10. This confirms that any uncertainty related to multicollinearity will not affect the results.

Results

Hedonic model analysis

In this section we utilise the database to construct the hedonic model for the office building transactional price index. Table 6.3 exhibits the regression results using the hedonic model.

TABLE 6.1 Summary of data characteristics

Variable	Description	Mean	Std. Dev.	Min	Max	p25	p50	p75
Yearbuilt	Year the property was built	1961	33	1840	2015	1939	1972	1986
SalepriceM	Transactional price in US$ million	48	76	5	675	9	22	54
Size	Size of the traded property	12667	14549	258	125552	3483	7208	18050
Landarea	Land size of the traded property in m²	1775	1782	113	13532	590	1055	2330
Year of sale	Year the property was sold	2007	5	2000	2015	2003	2007	2012
Capital value	Transitional price per m²	3993	2687	381	24131	2219	3304	4966
lncapital	Transitional price per m² (log)	8.11	0.62	5.94	10.09	7.70	8.10	8.51
Age	The difference between year built and date of sale	55.18	32.69	1.00	176.00	30.00	44.00	77.00
lnsize	Area of the office property (log)	8.86	1.17	5.55	11.74	8.16	8.88	9.80
lnlandarea	Land size of the traded office property (log)	7.07	0.91	4.73	9.51	6.38	6.96	7.75
lndisthigh	Distance to the traded office property with highest price per m² (log)	-4.62	0.56	-7.99	-3.33	-4.71	-4.50	-4.30
lndistmc	Distance to Melbourne Central station (log)	-4.95	0.37	-6.68	-3.43	-5.16	-4.92	-4.72
lndistsc	Distance to Southern Cross station (log)	-4.76	0.57	-6.47	-3.58	-5.09	-4.78	-4.39
lndistfs	Distance to Flinders Sreet station (log)	-4.93	0.47	-6.50	-3.25	-5.14	-4.83	-4.61
Dgrade1	Building has 'A grade' rating	0.16	0.36	0.00	1.00			
Dgrade2	Building has 'B grade' rating	0.47	0.50	0.00	1.00			
Dgrade3	Building has 'C grade' rating	0.25	0.44	0.00	1.00			
Dgrade4	Building has 'D grade' rating	0.08	0.27	0.00	1.00			
Dgrade5	Building has 'P grade' rating	0.03	0.16	0.00	1.00			
Dgrade6	Building has 'S grade' rating	0.02	7.55	0.00	1.00			
Dyearsale1	Transaction recorded in 2000	0.06	3.89	0.00	1.00			
Dyearsale2	Transaction recorded in 2001	0.10	0.31	0.00	1.00			
Dyearsale3	Transaction recorded in 2002	0.06	0.24	0.00	1.00			

(continued)

TABLE 6.1 (Cont.)

Variable	Description	Mean	Std. Dev.	Min	Max	p25	p50	p75
Dyearsale4	Transaction recorded in 2003	0.08	0.27	0.00	1.00			
Dyearsale5	Transaction recorded in 2004	0.07	0.25	0.00	1.00			
Dyearsale6	Transaction recorded in 2005	0.05	0.22	0.00	1.00			
Dyearsale7	Transaction recorded in 2006	0.06	0.24	0.00	1.00			
Dyearsale8	Transaction recorded in 2007	0.07	0.25	0.00	1.00			
Dyearsale9	Transaction recorded in 2008	0.05	0.22	0.00	1.00			
Dyearsale10	Transaction recorded in 2009	0.04	0.21	0.00	1.00			
Dyearsale11	Transaction recorded in 2010	0.08	0.27	0.00	1.00			
Dyearsale12	Transaction recorded in 2011	0.03	0.16	0.00	1.00			
Dyearsale13	Transaction recorded in 2012	0.03	0.16	0.00	1.00			
Dyearsale14	Transaction recorded in 2013	0.06	0.24	0.00	1.00			
Dyearsale15	Transaction recorded in 2014	0.07	0.26	0.00	1.00			
Dyearsale16	Transaction recorded in 2015	0.09	0.29	0.00	1.00			

The estimated hedonic model in Table 6.3 has been sufficiently robust to control for the possible impact of heteroscedasticity. As shown in Table 6.3, the regression model utilises 289 observations which explained approximately 70% of the office buildings' transactional price. From a spatial perspective there was a negative correlation observed between (a) office property with the highest transactional price per m² and (b) distance to important locations such as Flinders Street railway station. It was noted that the office property transactional price per m² is negatively correlated with the size and the age of the building, a finding consistent with the previous research. The transfer price of office property with an A or P (Premium) grade is substantially higher than the transfer price of office property with a lower grade. The coefficients for all annual dummy variables were positive and significant, therefore confirming the strength of the office property transactional price index based on these coefficients.

GIS spatial analysis

The plan of the Melbourne CBD is shown in Figure 6.1, which highlights the grid nature of the city. Of relevance to the research is the location of the three railway stations on the northern, southern and western sides of the precinct. The location and spatial proximity of individual properties to each other is important in explaining the transactional price of the office properties. Some of this information has been captured by the variables "lndisthigh", "lndistmc", "lndistsc" and "lndistfs" as shown in Table 6.1, which measures the distances between sold office property with (a) the highest price per m² and (b) Melbourne Central train station, Southern Cross train station and Flinders Street train station.

In addition to the distance to these key locations an Inverse Distance Weights Matrix was constructed to measure the proximity of the observations to each other. This Inverse Distance Weights Matrix will also be used in the test of spatial autocorrelation and the construction of spatial lagged and spatial error models which can control for the spatial dependency. This Inverse Distance Weights Matrix is calculated based on the latitude and longitude of each observation, and its statistical information is summarised in Table 6.4.

The total number of observations is 289, so the next step was to calculate 289 distances with other observations for each observation (including the distance with itself which is 0) in the matrix. As a result, the dimension of this matrix is 289x289. The parameter is set as "1" unit of latitude or longitude, which is approximately equivalent to 78km and much larger than the maximum distance between observations, which is 0.044 unit of latitude or longitude (3.432km). Thus, the distances with all other observations are measured for each observation in the matrix.

The spatial lagged model and spatial error model are estimated by using the above spatial weights matrix, and compared to the results of the hedonic model. The results of spatial lagged model and spatial error model are exhibited in Table 6.5. The index based on the spatial error model and spatial lagged model are used to compare with the index estimated by the basic hedonic model in Figure 6.2.

As Figure 6.2 shows, the office property transactional price increased by approximately 160% according to spatial error and spatial lagged model and approximately 140% according to hedonic model from 2000 to 2015. The office property price in Melbourne decreased after the 2007 global financial crisis, then commenced a recovery in 2009 and continued increasing until 2017. Note that the spatial error model and spatial lagged model produced

TABLE 6.2 Correlation coefficients

		(1)	(2)	(3)	(4)	(5)	(6)	(7)	(8)	(9)	(10)	(11)	(12)
Age	(1)	1.00											
lnSize	(2)	-0.52	1.00										
lnlandS	(3)	-0.38	0.68	1.00									
Dgrade1	(4)	-0.29	0.49	0.46	1.00								
Dgrade2	(5)	-0.18	0.12	0.08	-0.40	1.00							
Dgrade3	(6)	0.28	-0.39	-0.37	-0.25	-0.55	1.00						
Dgrade4	(7)	0.23	-0.29	-0.23	-0.12	-0.27	-0.17	1.00					
Dgrade5	(8)	-0.11	0.20	0.23	-0.07	-0.16	-0.10	-0.05	1.00				
DyearS2	(9)	0.02	0.02	-0.01	-0.05	0.04	0.01	-0.01	0.01	1.00			
DyearS3	(10)	-0.03	0.10	0.12	-0.03	0.03	-0.04	0.04	0.05	-0.09	1.00		
DyearS4	(11)	0.05	-0.04	-0.02	-0.05	0.10	-0.02	-0.03	-0.05	-0.10	-0.07	1.00	
DyearS5	(12)	-0.08	0.06	0.05	-0.04	0.00	0.01	0.03	0.04	-0.09	-0.07	-0.08	1.00
DyearS6	(13)	-0.06	0.02	0.05	0.11	-0.10	-0.03	0.11	-0.04	-0.08	-0.06	-0.07	-0.06
DyearS7	(14)	-0.04	0.02	0.03	0.09	-0.04	0.01	-0.02	-0.04	-0.09	-0.06	-0.07	-0.07
DyearS8	(15)	0.04	-0.08	-0.12	-0.04	-0.08	0.14	0.03	-0.04	-0.09	-0.07	-0.08	-0.07
DyearS9	(16)	0.00	-0.11	-0.08	-0.01	-0.03	0.04	0.05	-0.04	-0.08	-0.06	-0.07	-0.06
DyearS10	(17)	-0.08	0.06	0.03	0.00	0.00	-0.05	0.00	0.17	-0.07	-0.05	-0.06	-0.06
DyearS11	(18)	0.10	-0.07	-0.06	-0.05	0.07	-0.02	-0.08	0.03	-0.10	-0.07	-0.08	-0.08
DyearS12	(19)	-0.03	0.07	0.07	0.04	0.01	0.00	-0.05	-0.03	-0.06	-0.04	-0.05	-0.04
DyearS13	(20)	0.04	-0.07	0.02	-0.07	0.01	0.05	-0.05	-0.03	-0.06	-0.04	-0.05	-0.04
DyearS14	(21)	-0.07	0.10	0.10	0.09	-0.04	-0.08	-0.07	0.04	-0.09	-0.06	-0.07	-0.07
DyearS15	(22)	0.10	-0.07	-0.11	-0.01	-0.08	0.11	0.02	-0.05	-0.10	-0.07	-0.08	-0.07
DyearS16	(23)	0.02	-0.04	-0.02	0.03	0.04	-0.04	-0.04	0.02	-0.11	-0.08	-0.09	-0.08
lnDistHigh	(24)	0.01	-0.02	-0.06	-0.23	0.18	0.01	0.07	-0.04	0.10	-0.06	0.05	-0.02
lnDistMC	(25)	-0.13	0.18	0.16	0.07	0.08	-0.12	-0.03	0.02	-0.06	0.06	0.02	0.05
lnDistSC	(26)	0.04	-0.05	-0.07	0.10	-0.18	0.13	-0.09	0.08	-0.05	0.03	0.00	-0.01
lnDistFS	(27)	-0.30	0.11	0.21	0.16	0.06	-0.26	-0.01	0.09	-0.07	0.06	-0.07	0.04
VIF		1.79	6.95	5.43	12.46	1.06	5.15	6.25	3.41	2.45	1.89	2.11	1.96
Mean VIF		2.95											

very similar indexes. The index from the hedonic model, although with a very similar trend, exhibited higher volatility especially between the years 2007–09, which is the period incorporating the 2007 global financial crisis. It confirms that including the spatial weights matrix to control for the spatial dependency does change the estimated office property price index. This result is consistent with the estimated coefficients of "lambda" and "rho", which are positive and significant at 1% level in spatial error model and spatial lagged model. The "Wald test", "Likelihood ratio test", and "Lagrange multiplier test" following the spatial error model and spatial lagged model further support the finding that the "lambda" and "rho" play a significant role in explaining the office property transactional price. These results confirm the market price of office property is positively correlated with the market price of the nearby office properties. The shorter geographic distance between the office properties, then the more similar transitional price can be found. It indicates the existence of spatial dependency in explaining the office property transactional price. Furthermore, we employ the Moran's I and Lagrange multiplier to provide a more official and robustness test for the spatial dependency. Table 6.6 presents the results from these tests.

(13)	(14)	(15)	(16)	(17)	(18)	(19)	(20)	(21)	(22)	(23)	(24)	(25)	(26)	(27)
1.00														
-0.06	1.00													
-0.06	-0.07	1.00												
-0.05	-0.06	-0.06	1.00											
-0.05	-0.06	-0.06	-0.05	1.00										
-0.07	-0.07	-0.08	-0.07	-0.06	1.00									
-0.04	-0.04	-0.04	-0.04	-0.04	-0.05	1.00								
-0.04	-0.04	-0.04	-0.04	-0.04	-0.05	-0.03	1.00							
-0.06	-0.07	-0.07	-0.06	-0.06	-0.07	-0.04	-0.04	1.00						
-0.07	-0.07	-0.07	-0.07	-0.06	-0.08	-0.05	-0.05	-0.07	1.00					
-0.07	-0.08	-0.08	-0.07	-0.07	-0.09	-0.05	-0.05	-0.08	-0.09	1.00				
0.01	-0.05	-0.06	0.04	-0.01	0.03	0.03	-0.11	-0.01	-0.04	0.04	1.00			
-0.04	-0.11	-0.04	0.01	0.01	0.04	0.03	-0.02	0.07	-0.04	0.03	0.45	1.00		
0.01	0.04	0.09	-0.06	0.09	-0.07	-0.03	0.01	0.01	0.02	-0.05	-0.65	-0.35	1.00	
0.10	-0.02	-0.17	0.04	0.00	0.05	0.02	0.03	0.07	-0.06	0.06	0.09	0.22	- 0.40	1.00
1.82	1.94	1.98	1.8	1.71	2.16	1.43	1.53	2.06	2.07	2.31	2.45	1.43	2.36	1.72

Each analysis that used Moran's I and a Lagrange multiplier employed the same dependent and independent variables listed in Table 6.4. With reference to the results from the Moran's I and Lagrange multiplier tests, the spatial dependency (autocorrelation) was observed for the transfer price of office property. Note that this spatial dependency cannot be explained by the physical characteristics, transaction time or the distance to some key locations in the hedonic price model; it can only be captured by spatial weights matrix in the spatial error and spatial lagged model. Furthermore, we employed a root mean square error (RMSE) test to investigate whether including spatial weights matrix to construct spatial lagged and spatial error models can produce a more consistent and accurate estimation of office price indexes than the less complex hedonic price model. Table 6.7 summarises the test results.

To perform the RMSE test we randomly selected 80% (231) of the observations from the total database. We then estimated the new coefficients using a hedonic price model, spatial error model and spatial lagged model based on the newly selected sub-database. In the next step the new estimated coefficients were used to predict the residuals for the remaining 20% (58) observations and then we calculated their deviation (RMSE). We repeated this procedure five

TABLE 6.3 Basic hedonic model summary

Independent variables	Dependent variable: transactional price per m^2
	Hedonic model
age	-4.17E-04 (-0.50)
lnSize	-0.22★★★ (-5.85)
lnlandsize	0.11 (1.61)
Dgrade1	0.83★★★ (4.52)
Dgrade2	0.39 (1.71)
Dgrade3	0.12 (0.57)
Dgrade4	-0.20 (-0.55)
Dgrade5	1.14★★★ (4.99)
DyearSale2	0.26★★★ (3.49)
DyearSale3	0.48★★★ (4.32)
DyearSale4	0.45★★★ (3.76)
DyearSale5	0.46★★★ (4.95)
DyearSale6	0.88★★★ (5.99)
DyearSale7	0.92★★★ (7.19)
DyearSale8	1.18★★★ (9.40)
DyearSale9	1.14★★★ (8.55)
DyearSale10	0.83★★★ (7.31)
DyearSale11	0.95★★★ (8.53)
DyearSale12	1.15★★★ (7.18)
DyearSale13	1.18★★★ (8.44)
DyearSale14	1.24★★★ (10.35)
DyearSale15	1.40★★★ (12.75)
DyearSale16	1.45★★★ (14.50)
lnDistHighest	-0.02 (-0.37)
lnDistMC	0.08 (1.29)
lnDistSC	0.02 (0.39)
lnDistFS	-0.11★★ (-1.96)
_cons	8.61★★★ (12.64)
Obs	289
Prob > F	0.00
R-squared	0.70

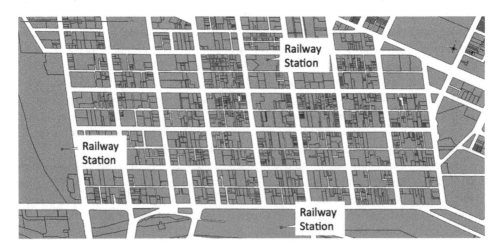

FIGURE 6.1 Location of Melbourne CBD grid and railway stations

TABLE 6.4 Summary of the inverse distance weights matrix

Dimension:	289x289		Minimum distance:	0	
Distance band:	0 < d <= 1		1st quartile distance:	0.004	0.312km
Friction parameter (d):	1	78km	Median distance:	0.006	0.468km
Largest minimum distance:	0.0234	1.8252km	3rd quartile distance:	0.01	0.78km
Smallest maximum distance:	0.0235	1.8252km	Maximum distance:	0.044	3.432km

TABLE 6.5 Hedonic model, spatial error model and spatial lagged model

Independent variables	Dependent variable: Transactional price per m^2	
	Spatial error model	Spatial lagged model
age	-4.04E-04	-4.02E-04
	(-0.52)	(-0.52)
lnSize	-0.27★★★	-0.26★★★
	(-6.15)	(-6.02)
lnlandsize	0.09	0.08
	(1.73)	(1.58)
Dgrade1	0.84★★★	0.82★★★
	(4.43)	(4.38)
Dgrade2	0.27	0.25
	(1.50)	(1.45)
Dgrade3	0.06	0.05
	(0.36)	(0.29)
Dgrade4	-0.14	-0.16
	(-0.76)	(-0.86)
Dgrade5	1.08★★★	1.07★★★
	(4.90)	(4.92)
DyearSale2	0.34★★★	0.36★★★
	(3.57)	(3.67)
DyearSale3	0.46★★★	0.48★★★
	(4.24)	(4.33)
DyearSale4	0.42★★★	0.42★★★
	(3.93)	(4.04)
DyearSale5	0.54★★★	0.55★★★
	(5.11)	(5.12)
DyearSale6	0.72★★★	0.71★★★
	(6.10)	(6.14)
DyearSale7	0.83★★★	0.84★★★
	(7.46)	(7.64)
DyearSale8	1.07★★★	1.09★★★
	(9.98)	(10.06)
DyearSale9	1.02★★★	1.03★★★
	(8.77)	(8.94)
DyearSale10	0.91★★★	0.93★★★
	(7.61)	(7.77)
DyearSale11	0.95★★★	0.94★★★
	(8.85)	(8.91)

(continued)

TABLE 6.5 (Cont.)

Independent variables	Dependent variable: Transactional price per m²	
	Spatial error model	Spatial lagged model
DyearSale12	1.07***	1.07***
	(7.69)	(7.73)
DyearSale13	1.21***	1.28***
	(8.82)	(8.97)
DyearSale14	1.27***	1.24***
	(10.65)	(10.96)
DyearSale15	1.43***	1.41***
	(13.22)	(13.36)
DyearSale16	1.53***	1.54***
	(15.09)	(15.28)
lnDistHighest	-0.01	0.00
	(-0.26)	(0.02)
lnDistMC	0.07	0.05
	(1.03)	(0.75)
lnDistSC	0.02	-0.01
	(0.33)	(-0.10)
lnDistFS	-0.11*	-0.13**
	(-1.90)	(-2.35)
_cons	8.59***	5.10***
	(12.84)	(2.57)
lambda	1.06**	
	(2.01)	
rho		1.82***
		(2.72)
Obs	289	289
Prob > F		
R-squared		
Squared corr.	0.72	0.72
Sigma	0.32	0.32
Log likelihood	-85.09	-83.90
Wald test of lambda=0:	4.23 (0.01)	
Likelihood ratio test of lambda=0:	3.76 (0.01)	
Lagrange multiplier test of lambda=0:	5.89 (0.00)	
Wald test of rho=0:		6.463 (0.00)
Likelihood ratio test of rho=0:		6.207 (0.00)
Lagrange multiplier test of rho=0:		6.258 (0.00)

times, where the RMSE for each model in Table 6.7 is the average root mean square error of the five results. As Table 6.7 shows, the spatial error model and spatial lagged model produced a very similar accuracy of the estimation of the coefficients and price index, which is superior than the hedonic model without controlling for the spatial dependency. Therefore, we can conclude that including the spatial dependency by using a spatial error model and a spatial lagged model can improve the accuracy of the office property price index.

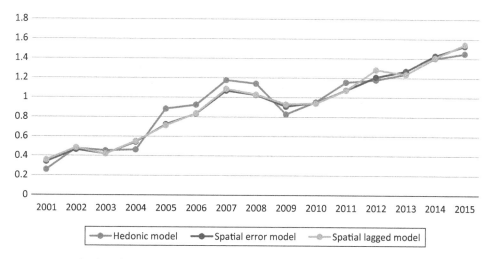

FIGURE 6.2 Index based on hedonic model, spatial error model and spatial lagged model

TABLE 6.6 Moran's I and Lagrange multiplier

Test	Statistic	df	p-value
Spatial error model:			
Moran's I	2.927	1	0.003
Lagrange multiplier	3.89	1	0.04
Robust Lagrange multiplier	3.78	1	0.05
Spatial lag model:			
Lagrange multiplier	6.213	1	0.013
Robust Lagrange multiplier	4.156	1	0.036

TABLE 6.7 Root Mean Square Error (RMSE) test

	Hedonic model	Spatial error model	Spatial lagged model
RMSE	0.76	0.65	0.66
RMSE/mean of "lncapital"	0.1	0.07	0.07

Conclusion

This was an innovative study which used spatial data to conduct a transactional price index for office property buildings in the Melbourne CBD which sold between 2000 and 2015. We found that the issue of spatial dependency existed for the transactional market of office buildings. It was concluded that the spatial error model and spatial lagged model produced very similar indexes. Furthermore, the index from the hedonic model, although with a very similar trend, exhibited high volatility especially between the years 2007–2009 which is the period incorporating the 2007 global financial crisis. This finding confirms that including the spatial weights matrix to control for the spatial dependency changes the output for the office property price index.

The findings of this research are consistent with the previous research that examined the office property unit transactions which are different from the office building transactions. Findings from this research contributed to the real estate and spatial sciences literature by highlighting the importance of spatial dependency in the valuation of office property buildings. Future research can be extended to test the spatial dependency in other aspects of the office property market, such as yield and rental income. Future research is encouraged to replicate this study in other global cities including London, New York and Hong Kong, whose markets are sufficiently large to produce data relating to whole building transactions. Also, it would be advantageous to extend the timeline to examine variations (if any) in the spatial weights over an extended time period.

References

Andersson, M. and Gråsjö, U. (2009). Spatial dependence and the representation of space in empirical models. *The Annals of Regional Science*, 43(1), 159–180.

Bollinger, C.R., Ihlanfeldt, K.R. and Bowes, D.R. (1998). Spatial variation in office rents within the Atlanta region. *Urban Studies*, 35(7), 1097–1118.

Bonde, M. and Song, H.S. (2013). Is energy performance capitalized in office building appraisals? *Property Management*, 31(3), 200–215.

Dermisi, S. and McDonald, J. (2010). Selling prices/sq. ft. of office buildings in down town Chicago – how much is it worth to be an old but class a building? *Journal of Real Estate Research*, 32(1), 1–21.

Dubin, R.A. (1992). Spatial autocorrelation and neighborhood quality. *Regional Science and Urban Economics*, 22(3), 433–452.

Dunse, N., Leishman, C. and Watkins, C. (2001). Classifying office submarkets. *Journal of Property Investment and Finance*, 19(3), 236–250.

Eichholtz, P., Kok, N. and Quigley, J.M. (2010). Doing well by doing good? Green office buildings. *The American Economic Review*, 100(5), 2492–2509.

Ericson, L.E., Song, H.S., Winstrand, J. and Wilhelmsson, M. (2013). Regional house price index construction – the case of Sweden. *International Journal of Strategic Property Management*, 17(3), 278–304.

Fuerst, F. and McAllister, P. (2011). Green noise or green value? Measuring the effects of environmental certification on office values. *Real Estate Economics*, 39(1), 45–69.

Henneberry, J. and Mouzakis, F. (2014). Familiarity and the determination of yields for regional office property investments in the UK. *Regional Studies*, 48(3), 530–546.

Huang, H., Miller, G.Y., Sherrick, B.J. and Gomez, M.I. (2006). Factors influencing Illinois farmland values. *American Journal of Agricultural Economics*, 88(2), 458–470.

Kim, H.M., O'Connor, K.B. and Han, S.S. (2015). The spatial characteristics of global property investment in Seoul: A case study of the office market. *Progress in Planning*, 97, 1–42.

Jones, C. and Orr, A. (2004). Spatial economic change and long-term urban office rental trends. *Regional Studies*, 38(3), 281–292.

Lee, M.L. and Pace, R.K. (2005). Spatial distribution of retail sales. *The Journal of Real Estate Finance and Economics*, 31(1), 53–69.

Liang, J. and Wilhelmsson, M. (2011). The value of retail rents with regression models: a case study of Shanghai. *Journal of Property Investment and Finance*, 29(6), 630–643.

Lizieri, C. and Pain, K. (2014). International office investment in global cities: the production of financial space and systemic risk. *Regional Studies*, 48(3), 439–455.

Maury, T.P. (2009). A spatiotemporal autoregressive price index for the Paris office property market. *Real Estate Economics*, 37(2), 305–340.

Nappi-Choulet, I., Maleyre, I. and Maury, T.P. (2007). A hedonic model of office prices in Paris and its immediate suburbs. *Journal of Property Research*, 24(3), 241–263.

Nitsch, H. (2006). Pricing location: a case study of the Munich office market. *Journal of Property Research*, 23(2), 93–107.

Reed, R.G. (2015). *The Valuation of Real Estate*, edited by R. Reed, Australian Property Institute, Canberra.

Shi, S., Young, M. and Hargreaves, B. (2009). Issues in measuring a monthly house price index in New Zealand. *Journal of Housing Economics*, 18(4), 336–350.

Shiller, R.J. (1991). Arithmetic repeat sales price estimators. *Journal of Housing Economics*, 1(1), 110–126.

Sivitanidou, R. (1995). Urban spatial variations in office-commercial rents: The role of spatial amenities and commercial zoning. *Journal of Urban Economics*, 38(1), 23–49.

Smith, S., Woodward, L. and Schulman, C. (2000). The effect of the Tax Reform Act of 1986 and over-built markets on commercial office property values. *Journal of Real Estate Research*, 19(3), 301–320.

Tobler, W.R. (1970). A computer movie simulating urban growth in the Detroit region. *Economic Geography*, 46(sup 1), 234–240.

Tu, Y., Yu, S.M. and Sun, H. (2004). Transaction-Based Office Price Indexes: A Spatiotemporal Modeling Approach. *Real Estate Economics*, 32(2), 297–328.

7

AN AGENT-BASED MODEL FOR HIGH-DENSITY URBAN REDEVELOPMENT UNDER VARIED MARKET AND PLANNING CONTEXTS

Simone Zarpelon Leao, Benoit Gaudou and Chris Pettit

Introduction

Compact city policies implemented through mixed-use higher density urban renewal and infill development are reshaping cities worldwide (Randolph 2006; OECD 2012; UN Habitat 2013). In the past, most of the urban growth occurred at the fringes of cities through expansion and sprawl. Also, we have seen brownfield development reconverted into housing estates after the relocation of manufacturing and industrial activities to the outskirts of the city, as driven through land value. As the inner and middle suburbs become increasingly built up with housing stock, and urban sprawl is understood to be inefficient with respect to infrastructure costs, there is increasing attention directed towards areas of existing housing to be renewed to provide additional dwellings for a growing population (Pinnegar *et al.* 2015).

Figure 7.1 illustrates the context faced by many urban planners in relation to high-density urban redevelopment. Figure 7.1.a shows that, theoretically, all the space up to the maximum building height permitted by the planning framework (controls) can potentially be redeveloped into new build stock. In reality, however, only part of that space is 'eligible' for redevelopment every year (Figure 7.1.b). The eligibility depends on the age of the existing building, whether it is for sale, and if it is economically feasible to redevelop the site (assuming a minimum profit margin and the existing planning framework). Redevelopment projects can be proposed within the current planning maximum height, or above that (Figure 7.1.c). In the State of New South Wales in Australia, the latter type of proposal, if refused by the local urban planning agency, can be taken by the developer to the State Joint Regional Planning Panel (JRPP), where developers and planners have the opportunity to defend their positions, and the panel decides the final outcome.

This tension is in part caused by the different outcomes which urban-planners and land-developers work towards. Urban-planners are driven by the desire to create functioning precincts with the vision of creating sustainable and liveable communities, whilst land-developers tend to consider profitability as the primary outcome. Since urban planning endeavours to respond to housing future demands, promote the revitalisation of areas and activate economic growth through redevelopment, careful consideration is given to height

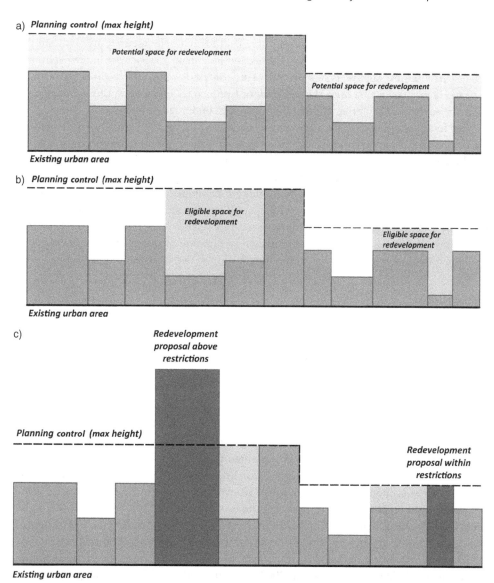

FIGURE 7.1 Potential, eligible and proposed space for urban redevelopment: hypothetical example of a precinct skyline: a) Potential total space for urban redevelopment according to existing built form and planning control; b) Eligible space for urban redevelopment per year considering existing built form, planning controls, age of buildings, whether a parcel is for sale and economic feasibility for redevelopment under existing planning controls; c) Redevelopment proposals within and above planning controls

envelopes. These height controls are enforced through zoning provisions. A development proposal which is significantly above the stipulated planning controls in a residential area is likely to have adverse impacts on the existing neighbourhood amenity and thus is given careful consideration by local planners. If such a proposal is approved, it also sets a precedent for further development of this type before the area is ready for such high density. On the other hand, if a development significantly above planning controls is proposed in an area with scarce or no availability of suitable and economically feasible land for redevelopment, it may indicate the need for a revision of current planning controls, and potentially the increase in the maximum height permitted in some parts of a precinct. This context is very relevant for the profession of property and real estate, since the interplay of land market forces and planning regulation affect the supply and the price of properties in an area under redevelopment.

This research argues that the decision on whether approval for a proposal is granted or not if it exceeds planning controls, and the revision of planning control parameters, should be based on analytical processes integrating information on the existing built form, land market and planning controls; these should provide an assessment of how much redevelopment an area can promote which is economically feasible for developers, and if this potential redevelopment meets the demand for housing in the area. There is currently a paucity of this type of analytical tool which can support urban-planners to evaluate redevelopment options. It is in this context that planning will respond to individual pressures from spot rezonings, which can have a negative effect on the amenity and functionality of precincts and municipalities.

Background

Previous research has brought to light relevant dimensions associated with the urban redevelopment process, and also presented some analytical methodologies to investigate urban redevelopment dynamics. Helms (2003) focused his study on factors affecting urban rehabilitation. He developed an empirical analysis of determinants of housing renovation in areas experiencing gentrification. Although this is not the same as urban redevelopment, these two processes have a common goal of attracting investment to improve the built environment. One of the outcomes from this is an increase in higher density development. Helm found that age, size and number of units were significant factors associated to characteristics of the property being rehabilitated, and also indicated that zoning and planning characteristics, which are not included in his study, deserves further investigation. Troy *et al.* (2015a, 2015b) investigated the economic feasibility of urban redevelopment processes at a land parcel level. They developed empirical rules for assessing the economic feasibility of medium- and high-density urban redevelopment of multi-family properties across Greater Sydney in Australia. This research revealed a process strongly driven by the market with significant influence from the planning framework and neighbourhood characteristics. Their approach included similar factors previously identified by Helms (2003), with the addition of the planning framework as a key element of success. A limitation, however, of this approach is that the original model was non-spatial and static in nature.

Since the process of urban redevelopment is in fact spatial, Leao *et al.* (2017) progressed on the previous model, developing a version of Troy *et al.*'s model within the ESRI ArcGIS software. This new version of the model allows users, predominantly urban-planners, to geographically visualise and explore different scenarios of urban redevelopment according to varied decisions of planning controls, and also variations in the market conditions. The results from a survey with urban-planners testing the model in a participatory workshop indicated the model was considered a very useful planning support (PS) system tool (Leao *et al.* 2017).

Cities are temporal in nature, yet there is a paucity of models which explicitly take into account the dimension of time. A potential route to overcome the temporal limitation on urban land development models was proposed by Parker and Filatova (2008). They argue that land markets have unique features that make them appropriate for dynamic modelling techniques such as agent-based modelling. Specifically, they consider land markets as heterogenous commodity traded by heterogeneous agents; spatial and agent-agent interactions can be present; and they operate in non-equilibrium dynamics. With this is mind, they developed a conceptual design of an agent-based model of some aspects of urban economics, with focus on better understanding the interactions between multiple buyers and sellers and the results in land value. Later, Filatova et al. (2009) progressed on the former work and implemented and tested the conceptual framework into an agent-based model of urban land markets. Developed in the agent-based modelling Netlogo platform, the model used a hypothetical monocentric city represented as a generic grid, where various artificial buyer and seller agents negotiate land prices for residential purposes. Levy et al. (2013) developed this approach further, including developers as agents, and assessing outcomes of land markets associated to urban densification. This model also tested theoretical hypothesis of urban economics in an artificial and generic grid-based monocentric city. Although being rich platforms for scientists to test hypothesis of urban economics, the generic agent-based models described above have limited utility for capturing urban planning or real estate processes in real urban settings. There is a lack of agent-based models of urban land markets built on actual urban geographies and current economic and planning data which can or could be used for real world city planning decision-making.

With an increasing number of people moving to and inhabiting cities, urban-planners and urban professionals need new tools and methodologies to gain insights into ways to deliver effective responses to the push towards greater urban densification. This in turn has important considerations in the interplay between land and property markets and planning frameworks. This study describes the beginnings of an agent-based model that endeavours to address this challenge by assessing the economic feasibility of urban redevelopment within real world precincts at a parcel level. The model considers existing and potential built forms, and parameters associated to the current planning framework and land and property markets. The model also includes a visual portrayal in three dimensions (3D) of the complex dynamics of buildings redevelopment. Using a real case study, the Kensington-Kingsford Town Centres corridor in the south-east region of Sydney, Australia, the model performance was successfully evaluated in terms of its capacity to produce metrics and visualisations of potential scenarios of redevelopment.

The fundamental urban process being modelled and reported in this chapter is the vertical development of urban precincts from low- or medium- to high- or very high-density. In other words, detached houses or old multi-storey apartment buildings being knocked down and replaced by modern high-rise buildings. This chapter presents a novel approach to representing urban change in an agent-based modelling environment, in which land parcels are autonomous agents; they are fixed in space, as land parcels cannot move (i.e. they don't change their location), but they can change their height through redevelopment and densification. The parcels' behaviour is driven by ageing, suitability, availability, economic feasibility, planning compliance, and ultimately by replacement if a series of conditions are met. These conditions are dependent on the land parcels' initial conditions and some external parameters (defined by the market and urban planning). There are many agent-based urban models reported in the literature, but they are mostly generic and hypothetical (see for example Filatova et al. 2009 and Levy et al. 2013 described before). In instances that these models are more realistic, they have buildings as background display objects and the focus on the model is to understand patterns of mobility, whether it be people or vehicular movements. This is the first attempt at a novel approach to model the process of urban redevelopment using an agent-based modelling (ABM approach) with land parcels/buildings as autonomous agents. A 3D environment has been selected as the most suitable form of visualisation of the transformations simulated by the model. Moreover, as real world data can be available for precincts of interest, and as urban-planning agencies are increasingly using geographic information systems (GIS) and associated spatial data, this research has been undertaken in an advanced ABM platform which can handle such spatial data inputs.

Research questions and discussion

Within the context of the property and real estate profession and the specific case study analysed, three questions are posed:

(i) *What insights can the agent-based model of urban redevelopment provide to users?*
(ii) *Can the model be used to investigate potential future scenarios based on neighbourhood design proposals?*
(iii) *What are the main benefits and limitations of agent-based modelling for urban redevelopment related research?*

Research methods

Agent-based modelling (ABM) is a method under the umbrella of complexity. It is built upon the principle that emergence of macro-patterns comes from individualised but interconnected micro-decisions. The concept of 'micromotives and macrobehaviour' was stated in the homonymous book published by Thomas Schelling in 1978. However, it was the advances in computing that allowed its translation into sophisticated computer models (Crooks *et al.* 2008). The complexity inherent within cities is increasingly being addressed through ABM (Batty 2007). In ABM, autonomous agents interact with the environment and with each other through behavioural rules. Although the rules can be simple, the varied configurations of the environment and the agents may cause non-linear feedback, and result in surprising and unexpected results (Axelrod 2005). Agents can be people, vehicles, countries or urban parcels, as in the model presented here. Chen (2012) argues that the foundation and concepts of ABM are particularly suitable in socio-related studies, especially in urban studies and design project management.

This section of the chapter presents the proposed urban redevelopment ABM which follows the ODD protocol – Overview/Design Concepts/Details (Grimm *et al.* 2006). This protocol has been widely accepted and used as the standard to describe ABMs since its publication, and it has been confirmed as an appropriate protocol to support urban models (Polhill *et al.* 2008). Finally, in the chapter we describe the modelling software platform used to develop and test the model.

Overview

Purpose

The purpose of the model is to provide an understanding of how land and property markets (purchase cost, sale revenue, profit expectation) and urban-planning frameworks (buildings' maximum height and floor space ratio) affect high-density urban redevelopment (economic feasibility, regulation compliance, regulation revision and housing supply) at both the land parcel and precinct scale. The model has been developed and tested in part of the city of Sydney, NSW, Australia.

State variables and scales

The model works across two scales of geography (hierarchical levels): precinct and land parcel. Parcels are characterised by the state variables: area; type (built or vacant); age; whether it is a strata title; number of units/dwellings; current height; maximum height according to current planning framework. 'Strata title' allows individual ownership of part of a property (generally an apartment), combined with shared ownership in the remainder (common areas of the property, such as driveways, foyer, gardens, lifts, etc.) through a legal entity called the 'owners corporation'; it offers a legal mechanism for space to be vertically subdivided and traded, allowing individualised property rights to be applied to multi-unit housing (Troy *et al.* 2017). According to the current NSW Strata regulation, at least 75% of the units, through their owners' votes, are required to approve the sale of whole strata building for redevelopment (http://stratalaws.nsw.gov.au/).

Since parcels can be redeveloped (old building replaced by new building or vacant land filled with new building), its state variables (building age, height, number of units/dwellings and whether it is a strata, etc.) can change. A precinct, a higher-level entity, is formed by parcels (with or without buildings) with fixed area and location. Its properties are summations and averages of the attributes of its parcels: number of parcels (total, suitable, available, compliant, redeveloped), number of dwellings/units, average building height.

Each model run/step is equivalent to one year and the time horizon is theoretically unlimited, but it is constrained by a 'saturation' parameter (the default has been set at 90m building height for the study area, equivalent to approximately 30 floors). The model is designed for precinct level analytics, with land parcels as foundation spatial units. The data used is GIS-based, made available through the State Government Agency Property NSW, combined with other data sourced from state government and municipalities.

Process overview and scheduling

The parcels in the precinct can change their attributes as a result of two processes: ageing and redevelopment, the latter driven by the developer. Also, the planning framework may be revised by the urban planner according to criteria associated to land scarcity and density (Figure 7.2). The process of redevelopment contains a series of interlinked sub-processes: suitability, availability, amalgamation, economic feasibility and (re)placement. The planning framework encompasses sub-processes which influence redevelopment, including planning compliance, and framework revision.

Ageing: The age of a building is a criterion for its suitability to be redeveloped. Only buildings which are at least 25 years old are suitable for redevelopment. The age of an existing building is added one unit at the end of every model step, which represents a year. Every time a building is replaced by a new one as a result of redevelopment, its age is reset to zero, and the ageing process proceeds accordingly at the next step.

Redevelopment: As a result of the redevelopment assessment a parcel can be redeveloped. This process involves the replacement of an existing building by a new one, or 'placement' of a new building in

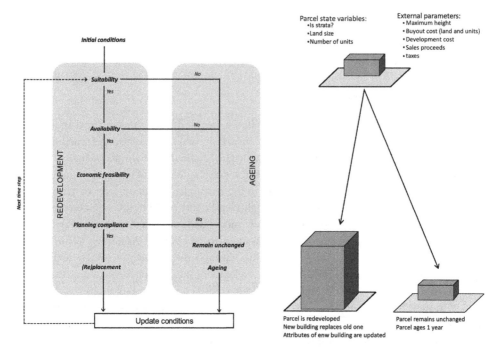

FIGURE 7.2 Simplified scheme of the agent-based model processes

a vacant land parcel; or the parcel remains unchanged. The redevelopment assessment is comprised of a series of stages which run at each time step of the model. They are described below:

- Age suitability: first, the suitability of a parcel for redevelopment is assessed based on its age; only parcels with buildings which are 25 years or older are considered 'suitable' for redevelopment. Vacant parcels are always suitable for redevelopment.
- Availability on the market: second, the availability of parcels previously classified as suitable is assessed. This process tries to emulate the fact that, although suitable, not all parcels would be available on the market for sale at a time. Based on a random number, parcels are attributed a binary state (available or not available).
- Economically feasible urban form: third, the model calculates the built form required for a new building to be economically feasible, considering a minimum profit required by the developer (by default it is 20%). This process is developed for the individual parcels previously classified as suitable and available. The calculation is calibrated to local land and property market conditions and construction costs at a precinct level. The new built form required for it to be economically feasible is described in terms of number of new dwellings/units and number of floors of the new building.
- Compliance to framework: fourth, the model compares the number of floors of the new building (to be economically feasible) to the maximum height permitted within the planning framework at the location (converted into number of floors). If the new height is below or equal to the maximum threshold, the building is considered 'compliant' and

the replacement of the old building can be performed in the next stage; otherwise, the old building remains unchanged at this time step.

- (Re)placement of old building or vacant land: fifth, if a parcel passed positively through all the previous assessments, being suitable, available, and compliant to the local planning framework, and at an economic feasibility height, it is listed as a candidate for redevelopment. Candidates will be selected and redeveloped within a model time step while the sum of new units provided by redevelopment minus the sum of units knocked down is below the annual housing demand for the area. The candidate parcels list is ranked in descending order of profit to development at maximum height, so the most profitable parcels available will be given preference. The attributes of the redeveloped parcels are updated with the characteristics of the new building.

After the processes of 'ageing unchanged parcels' and 'updating the attributes and age of new buildings', a new time step of the model is run again, and so forth, until the saturation parameter is reached.

Model design concepts

Emergence

Land redevelopment dynamics emerge from the parcels' behaviour, but their life cycle and behaviour are almost entirely represented by empirical rules describing, for example, suitability, economic feasibility, control compliance and replacement. Exception is for the availability, which is random. Urban redevelopment is emergent in the sense that it is the result of decentralised decisions of autonomous 'parcels'. Collectively these decisions will result in a housing supply (overall number of residential units) and an urban high-density landscape (number of floors of building by year in a 3D visualisation).

Sensing

In this first version of the model as reported in this chapter, for simplification purposes parcels are assumed to be autonomous agents, aware of their state attributes, aware of land market and planning conditions, able to evaluate their economic feasibility and compliance with planning, and capable of 'redevelopment' themselves if conditions are appropriate.

Interactions

The model does not consider the interplay between owners of old buildings/units, developers, buyers of new units and planners. This first version of the model considers only parcels to be agents. The assessment they perform on themselves to decide if they redevelop or not are through functions which include information from planners and developers. Moreover, parcels are assumed to be completely independent; the fact that a parcel is for sale or has been redeveloped recently does not affect the neighbouring cells. Indeed, the parcel-agent represents the behaviour of a developer willing to pay market price for a property and develop it attending to the planning controls if it can achieve at least 20% profit after sales of new units constructed. The model allows interaction with the user through the manipulation of model parameters associated to land and property markets, planning framework and some characteristics of new urban form.

Stochasticity

The availability of a parcel for sale at the land and property market at every single year (model step) is a random attribute for a parcel with low frequency (few parcels at a time). This means that a parcel for sale in one year may not be available the following year, even though it has not been sold and redeveloped. Also, a parcel for sale in the market, although available, may not be economically feasible for development at the planning controls at the time, and not be redeveloped for that reason. This means that even in a case study with real parcel attributes and calibrated empirical rules for land market and planning controls, each simulation will result in a different outcome, due to the stochasticity of the properties availability on the market, and the fact that all the subsequent assessments for urban redevelopment are dependent of the characteristics of the parcels available. This also means that, although parcels are agents 'aware' of influences by an economic and planning context (they represent the action of a developer), property sellers are not. This is considered appropriate, because not all parcel owners are necessarily potential sellers waiting for the best opportunity. Some may have long-term intentions to stay at their properties, or need to sell a property at a certain time regardless of market opportunities or adverse conditions.

Observations

Observations include a 3D visualisation of the parcels in a precinct. The model starts by showing current heights of buildings at time 0 of the simulation, and at each model time step parcels redeveloped are highlighted in colour (in comparison to the ones which remain unchanged) and also extruded in scale to the height of the new building. Within the ABM software platform used, graphs summarise some model outcomes: (a) number of suitable and redeveloped parcels in the precinct, and (b) average building heights vs. planning maximum height in the precinct.

Details

Initialisation

This model has been developed to work with actual city data. The main dataset is a GIS layer file with polygons representing parcels, with an associated attribute table containing the state variables of each parcel. The study area has around 1,500 parcels with varied sizes, age and density of development. Based on the current planning framework defined by the local government in the study area, maximum heights vary from three to ten floors, and floor space ratio (FSR) from 0.5 to 3, excluding the town centre area where FSR is not limited. Moreover, the model requires the user to input some parameters related to costs and revenues of new development. The cost of purchasing land and existing property that will be redeveloped was estimated based on NSW Valuer-General's sales data for properties in the local Census Statistic Area level 2 (SA2, Australian Bureau of Statistics) and the estimated stamp duty payable by a prospective developer. The cost of replacing existing stock with a new construction was estimated using the Rawlinsons Construction Cost Guide (Rawlinsons 2014). A cost per new dwelling was obtained for a two-bedroom unit with $90m^2$ and medium standard quality of construction based on the design guidelines proposed by the NSW State Environmental Planning Policy No. 65: Design Quality of Residential Apartment Development (SEPP 65). The revenue from selling the new property after redevelopment was estimated based on the current sales values of similar new built development in the SA2 area, discounted the 10% GST payable by the developer on sale.

For the case study, the estimated land market parameters used as default are: cost for purchasing an existing strata unit is A\$785,000; the cost for purchasing a vacant land or a non-strata property is A\$1,125/m^2; the construction cost of a predominant typical typology in the area is A\$215,000 (a two-bedroom apartment with 90m^2 and medium standard quality); and the sale price of a new redeveloped unit is A\$800,000. Note that the sale price of old and existing units is similar. The small difference between these two costs is overset by the fact that redevelopment produces a higher density new building with a larger number of units when compared to the old replaced buildings.

Additional default parameters used in the model include: the annual demand for new residential units is 200 units/year; site coverage is 60%; and the proportion of parcels for sale is 30%. Using an actual case study, the initial state variables are always the same. However, the random number used to define 'availability' on the market will make each model run a unique thread. Creation of scenarios is possible in the model by changing state variables in the attribute table of the parcel dataset, and also, from new values for model parameters (profit margin, land coverage, buyout cost, sale proceeds, annual housing demand, etc.).

Input

GIS feature data (polygon shapefile) with an attribute table describing associated state variables; parameters calibrated to a study area.

Sub-models

Sub-model 1 – suitability

'Suitability' of a parcel is tested at the start of every time step, as buildings are ageing (increased age) or being replaced (decreased age) during the urban redevelopment process. The parcels' type and age of an existing building are used as criteria for the suitability assessment, as shown in Figure 7.3.

The 'Renewing the Compact City Report' (Troy *et al.* 2015a) identified that a significant part of buildings actually redeveloped in Sydney were at least 35 years old. Moreover, when estimating the potential for redevelopment across Greater Sydney of properties built from 1990 (then 25 years old), the study also found a reasonable number of suitable properties for redevelopment based on their economic feasibility at maximum height restrictions. Therefore, being inclusive by using a minimum age requirement, this study selected 25 years as the default parameter for age suitability for redevelopment.

Sub-model 2 – availability

'Availability' state of a parcel is tested at the beginning of every time step, after sub-model 1. Whether a parcel is 'available' is defined through a random assignment with low frequency (tendency to have less available than unavailable parcels in the market) (Figure 7.4).

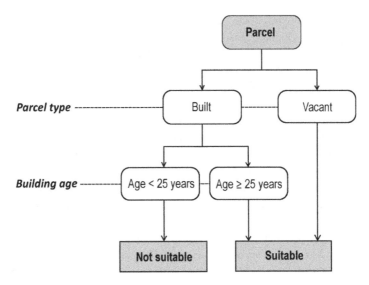

FIGURE 7.3 Suitability assessment for urban redevelopment

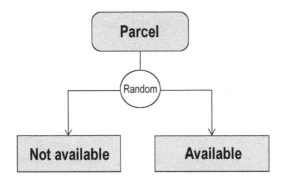

FIGURE 7.4 Availability assessment for urban redevelopment

Sub-model 3 – height of economically feasible building

For each parcel which is 'suitable' and 'available', the model calculates the number of units and number of floors for a new building which would need to be economically feasible to be redeveloped. The calculations evaluate all the costs involved in purchasing land (in the case of non-strata properties) or units of a building (in the case of strata properties), paying taxes, and constructing new units of a certain size, as well as all the revenues from their sales. To be economically feasible the new building must have a size which is able to provide the developer with a profit of at least 20% (all revenues minus all costs). The sequence of calculations and the description of the variables and constants used in the model are shown in Figure 7.5.

Sub-model 4 – compliance with planning controls

For each parcel which is 'suitable' and 'available', the model checks if a new building which is economically feasible complies or not with the planning framework. This assessment is

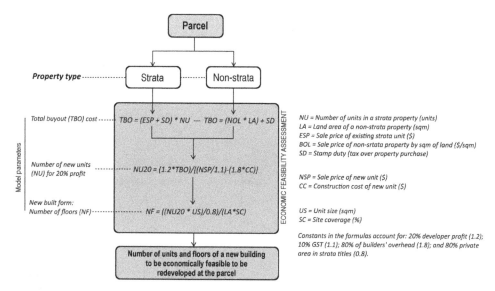

FIGURE 7.5 Economic feasibility assessment for urban redevelopment

made by comparing the number of floors of the new building with the maximum number of floors allowed at the parcel site. If the height of the new building is at or below the maximum threshold, the parcel is compliant, otherwise it is not compliant. This is illustrated in Figure 7.6.

Sub-model 5 – ageing

If a built parcel is 'not suitable', or 'not available', or 'not compliant', it remains unchanged and ages one year at the end of every time step:

Parcel age (built) = Parcel age (built) + 1

Sub-model 6 – building (re)placement

Redevelopment involves the replacement of old building by new building or placement of a new building on a vacant land. If a parcel is 'compliant', it will be listed as a candidate for redevelopment and all candidate parcels will be ranked in descending order of profitability of redevelopment at the maximum height. Parcels will be redeveloped at a model time step until the housing supplied by redevelopment minus the sum of units knocked down is equal to or higher than the annual demand of housing. The annual housing demand is an input provided by the user, and the supply is the accumulated number of units in the new redeveloped building at the minimum profitability. This process is illustrated in Figure 7.7. At the end of the time step, the state variables for the 'redeveloped' parcel are updated according to the characteristics of the new building defined in sub-model 3 calculations (number of units, number of floors, strata title) and age = 0. For simplification at this first version, it is assumed the new building is immediately built.

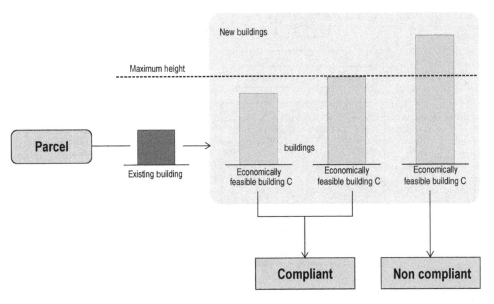

FIGURE 7.6 Planning compliance assessment for urban redevelopment

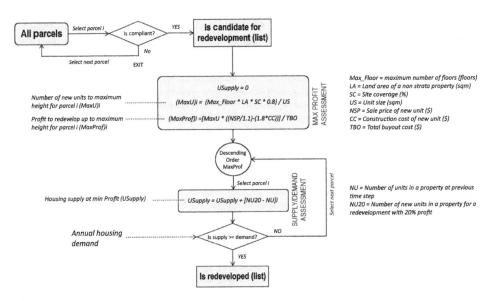

FIGURE 7.7 Selecting and redeveloping parcels

Modelling platform

The model has been developed using GAMA (www.gama–platform.org). This choice was based on: (a) the capability of GAMA in working with GIS data and instantiating agents from geographical features (Taillandier *et al.* 2010); (b) its good performance with 3D visualisation of GIS data (Grignard *et al.* 2013); and (c) for being a free, open source and well documented platform.

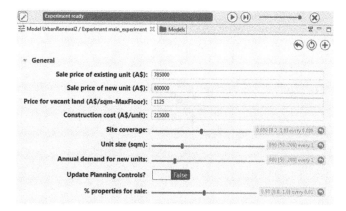

FIGURE 7.8 Model parameters for the study area

Results and analysis

To test the model performance, a case study has been selected and four experiments are developed: (a) simulation with past data and comparison with actual redevelopments occurred in the study area (validation); (b) simulation with present context to get insights into future redevelopments (simulation); (c) simulation for testing the economic feasibility of an actual design proposal for the study area; and (d) simulation of a simple automated trigger for revising planning framework.

Case study: location and parametrisation

The Kensington-Kingsford Town Centres corridor, in the south-east region of Sydney, Australia, is used as a case study in this research to test the use of the proposed agent-based model for economic feasibility assessment of urban redevelopment at the parcel level.

This is considered a rich case study due to two reasons. First, the planning agency of this region has been approached by several developers with requests for construction at building heights above thresholds established by the existing planning framework. These are expected to place pressures on urban growth and redevelopment in the area due to its favourable location, connectivity and high standard of services. Second, an urban design competition was launched in 2016 for this area, and the winning design proposed a significant reconfiguration for the area, without any explicit consideration for its economic feasibility (http://yoursayrandwick.com.au/k2k).

Figure 7.8 presents the model parameters input window, filled with estimated values as default for the study area to assess the economically feasibility of urban redevelopment in 2016. The parameters are provided in a simple graphical user interface for users to adjust and to run various *what if?* scenarios. The ability for end users to explore an envelope of *what if?* scenarios is powerful, as it enables various policy levers and planning instruments to be explored before they are set or, to assist in helping to review them (Pettit and Pullar 2004).

Validation

Data on nine new buildings that have been redeveloped on 30 parcels in the study area from 2009 to 2016 have been used to validate the model rationale and calculations. Figure 7.9.a

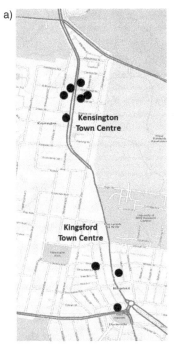

FIGURE 7.9 Study area and exemplar urban redevelopment: (a) Actual redevelopments in the study area 2009–2016; (b) Before: two parcels with a single-floor commercial unit each; (c) After: amalgamated parcels with a seven-floor building (42 dwellings), built from 2011–2013

locates those redevelopments in the study area and illustrates an urban redevelopment process in 2013, in which two parcels with single-floor commercial units (Figure 7.9.b) were amalgamated, and a seven-floor high-rise building was built as a replacement with 42 new residential units (Figure 7.9.c). It is tested here if the model would identify those areas as potential redevelopment sites.

The Urban Redevelopment ABM was run using 2009 as the starting year (parcels subdivision and existing built form before redevelopments) assuming that all parcels would be available in the market, and then the results were compared with the actual redevelopments which occurred in the area.

It was found that all the parcels redeveloped in the study area in reality were also virtually redeveloped by the ABM for multiple model runs. They were all considered suitable, economically feasible and compliant with the planning framework. Also, the model showed that those parcels produced the same profit for developers as the average of all suitable and compliant parcels in the area (42% profit if parcels were developed up to the maximum height).

However, the model could not replicate the actual timing and order of the redevelopments that occurred in the study area. This was expected, since the entry of a property in the market at every time step of the ABM is a random variable. Multiple runs of the model for the same period generate different outcomes at a parcel level, but similar outcomes when aggregated to the neighbourhood level. Therefore, the results of the model should not be taken as an accurate 'forecast' of urban redevelopment at a parcel level, but an insight into potential redevelopment and densification processes and patterns.

Simulations

Model simulations have been developed for four *what if?* scenarios: (1) business as usual; (2) an urban design proposal; (3) the urban design proposal with increased planning control; and (4) automated planning control revision. These are described below.

Business as usual scenario

Based on the conditions in the study area for the year 2016, this scenario is running a simulation of the model to answer the following questions: (1) How much of the area would be redeveloped if land/property market and planning control parameters remain as they currently are? (2) Would the area reach the full densification potential as stated by the planning controls (maximum height)?

Figure 7.10 illustrates the results of a simulation run for this business as usual (BAU) scenario. Multiple runs of the model resulted in comparable results. It indicates that most of the redevelopment would occur in the Kensington area (parcels in dark grey have been redeveloped, while light grey parcels remain unchanged, Figure 7.10a). This shows that Kingsford, in order to become economically feasible to be redeveloped, would need a higher allowance for building heights, floor space ratio (FSR) or higher sales prices for properties, or a combination of these. Indeed, the simulation indicates that in around 25 years there are no more eligible parcels for redevelopment (Figure 7.10b). Although the planning controls would allow further development, the lack of economic feasibility blocks redevelopment before reaching the full potential of the area. This indicates that a revision of planning controls for the Kingsford area may be required.

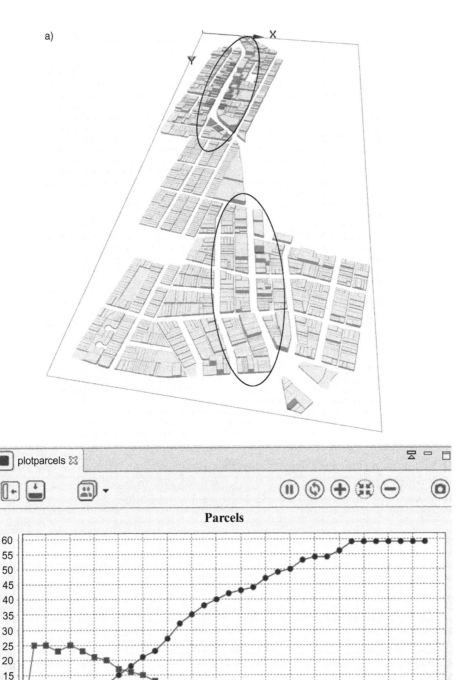

FIGURE 7.10 ABM for urban redevelopment for the business as usual scenario: a) 3D visualisation of redeveloped parcels; b) Eligible and redeveloped parcels annually

Simulation of a winning urban design proposal

The local government responsible for the study area launched an international urban design competition in 2016 to stimulate a future vision for the Kensington-Kingsford corridor region (www.youtube.com/watch?v=aOk7Aa-njAA). The winning design proposed significant changes to the local urban landscape and some changes to the current planning controls. These changes included the resurfacing of a creek and the creation of an extensive green open space corridor along this creek (outlined in black in Figure 7.11a). The reduction of the parcel sizes to accommodate the new green open space would be compensated by an increase in the building heights along the green corridor, which is an area currently occupied by low-density housing. Increased building heights was also envisioned for the main road corridor

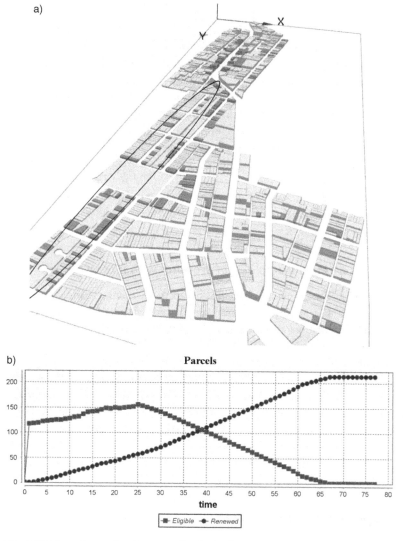

FIGURE 7.11 ABM for urban redevelopment for the 'Urban Design Proposal' scenario: a) 3D visualisation of redeveloped parcels; b) Eligible and redeveloped parcels annually

(Anzac Parade) linking the two town centres (Kensington and Kingsford, outlined in black in Figure 7.10). The design proposals for the competition had to respond to the overall goals of the local government in terms of the neighbourhood image, character, economic development, service provision and quality of life. The economic feasibility of the proposals was not assessed as part of the competition process. This, however, is an essential component if any design proposal aims to be delivered.

The simulations here are based on the 'desired future' proposed by the winner of the urban design competition in the study area. The new parcel sizes, types, and maximum planning controls were input into the ABM. Moreover, since the design proposal did not specify a FSR, the maximum current parameter used in the town centres was adopted in the green corridor, which is FSR 3:1. All areas not affected by the design proposal maintained their current characteristics. Land and property market parameters were considered the same as in the BAU scenario,

The model is used to answer the following questions: (1) *Is the urban design proposal economically feasible to be implemented?* (2) *How much of the area would be redeveloped in the conditions stated by the urban design proposal?* Figure 7.11 illustrates the results of a simulation for this second scenario. Multiple runs of the model resulted in similar results. When compared to the BAU scenario, it indicates that the proposal would significantly increase urban redevelopment in the area. It would be only after 65 years that eligible parcels would no longer be available in the area. The increased redevelopments are mostly from the new area zoned as high density along the green corridor (dark grey parcels), but it is noticeable that significant part of the parcels in this new zone are not economically feasible to be redeveloped (light grey parcels). This suggests that the design proposal would not be fully developed due to lack of economic feasibility, and that a revision of planning controls may be necessary to promote further redevelopment and densification of the area, as recognised in the design proposal.

Simulation of a winning urban design proposal with increased FSR

Based on the results of the previous simulations with the winning urban design proposal, the effect of increasing FSR from 3:1 to 5:1 on the urban redevelopment process in the study area is tested here. The simulations use the model to answer the following question: How much of the area would be redeveloped if the FSR is increased to 5:1 in high-density zones of the urban design proposal?

Figure 7.12 illustrates the results of a simulation for this third scenario. When compared to the previous scenario, it indicates that when FSR is increased, significant additional growth occurs along the new development area, as intended by the proposed design. This suggests the usefulness of the model to test design proposals in terms of their actual potential for implementation based on economic feasibility, which is in most cities driven strongly by the property market. The simulation results indicated that it is only from an FSR of 5:1 that the design proposal activates a majority of parcels as feasible.

Automated planning control revision

By default, the model runs progressively and at some point, when there are no more eligible parcels, the redevelopment process stops. However, the model also allows users to set

FIGURE 7.12 ABM for urban redevelopment for the 'Urban Design Proposal + increased FSR' scenario: a) 3D visualisation of redeveloped parcels; b) Eligible and redeveloped parcels annually

an automated planning control revision process. This is done by selecting "TRUE" for the 'Update Planning Controls?' in the initial input window of the model. When this is set, the model will update automatically the maximum building height when the eligibility is low. Eligibility is considered low when potential new dwellings in eligible parcels are 20% or less of the annual demand for new dwellings; and if this occurs, existing maximum heights in the area are increased in increments of 25% (i.e. a building with maximum height equivalent to four floors will be increased to five floors; and one with eight floors will be increased to ten floors). This set-up makes the system continuously increase maximum height in order to avoid

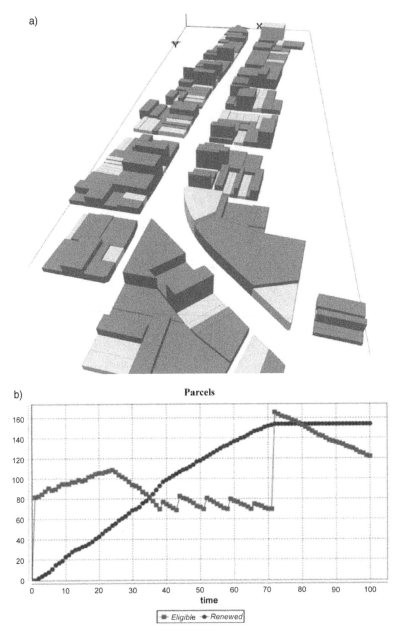

FIGURE 7.13 ABM for urban redevelopment with automated maximum height revision: a) 3D visualisation of redeveloped parcels; b) Eligible and redeveloped parcels annually

low eligibility. Although this is not a realistic representation of the planning controls' revision process, it assists in the visualisation of potential results of a market-driven process, in which planning controls would follow land market pressures in this 'build-out' scenario.

Figure 7.13 illustrates the results of a simulation for this build-out scenario, in which dark grey parcels have been redeveloped. The 'serrated' shape of the eligible parcels' graph indicates the timing of maximum height changes, and the consequent increase in parcels' availability at the new planning control parameter. This occurs because, with the new higher maximum height, some parcels become economically feasible with the additional units that can be built and sold. Currently, the model runs continually until a saturation height of 120m is achieved.

Conclusion

This chapter presented the first version of an agent-based model to assess and visualise the process of urban redevelopment of an area over time. The model is predominantly driven through the consideration of land market conditions and the current planning framework of a study area. One significant innovation in the proposed model is the fact that it works with actual geographies and disaggregated data at a parcel level, with a realistic neighbourhood visualisation. All the agent-based models currently reported in the literature dealing with the economics of urban redevelopment use artificial and simplified cities represented by generic grids. Despite being useful to test theoretical hypothesis, these generic and simplified models have very limited application in supporting planning and decision-making of areas under pressure for renewal in the real world.

The test of the model to a real case study demonstrated the strength of the model in (1) handling geographic data easily; (2) providing some meaningful insights of the combined effects of land market and planning framework on the urban redevelopment process; and (3) the potential of the model to be linked to urban design processes in order to assess their actual delivery potential based on economic feasibility.

In its first version, the model has some limitations which are important to note. Currently the parcel is an agent which is aware of the developer's goals and also aware of the restrictions imposed by the urban–planner at its location. The parcel reacts to these drivers. The intention here was to simplify as much as possible the variables involved in the system, while still keeping the core complexity. It is aligned with the understanding that planning regulations and market conditions are the main factors driving redevelopment in consolidated urban centres, and that their effects can be summarised at a parcel level. Further development could try to model the interaction of human agents involved in the urban redevelopment process, such as local residents, property owners, real estate agents, property buyers, investors, renters, banks/home-loaners, etc. These additional agents bring new goals, variables and a new level of complexity to the system. Another potential improvement is the capacity of the parcels to interact with each other, such as in an amalgamation process where the aggregation of two or more parcels may affect their capacity to be redeveloped. In future versions of the model this will be addressed through the development of spatial proximity and adjacency operations. Finally, the tests of the model reported here included a small neighbourhood, and its application to a large scale urban area will bring new challenges. One challenge is related to the performance of the platform in terms of computing time and visualisation to handle large scale contexts, particularly if more agents and interactions are added to the model. The second challenge is associated with the data requirements. We believe that advances in high-performance computation and the increasing availability of urban big data raise optimistic horizons for further development of realistic agent-based models to assist better understanding, planning and management of urban property development over the future.

Acknowledgements

The authors would like to thank Dr Laurence Troy and Professor Bill Randolph, from the City Futures Research Centre at the University of New South Wales, for sharing the empirical rules for urban redevelopment used in this research based on their previously published research, and the City of Randwick Council for providing data for the case study.

References

Axelrod, R. (2005). Advancing the art of simulation in the social sciences. In Rennard J-P (ed.), *Handbook of Research on Nature Inspired Computing for Economy and Management*, Hersey, PA: Idea Group.

Batty, M. (2007). *Cities and Complexity: Understanding Cities With Cellular Automata, Agent-Based Models, and Fractals*. The MIT Press.

Chen, L. (2012). Agent-based modelling in urban and architectural research: a brief literature review. *Frontiers of Architectural Research*, 1(2), 166–177.

Crooks, A., Castle, C. and Batty, M. (2008). Key challenges in agent-based modelling for geo-spatial simulation. *Computers, Environment and Urban Systems*, 32(6): 417–430.

Filatova, T., Parker, D. and van der Veen, A. (2009). Agent-based urban land markets: agent's pricing behaviour, land prices and urban land use change. *Journal of Artificial Societies and Social Simulation*, 12(1), 3. Retrieved from: http://jasss.soc.surrey.ac.uk/12/1/3.html.

Grimm, V., Berger, U., Bastiansen, F., Eliassen, S., Ginot, V., Giske, J., Goss-Custard, J., Grand, T., Heinz, S.K., Huse, G., Huth, A., Jepsen, J.U, Jorgensen, C., Mooij, W.M., Muller, B., Pe'er, G., Piou, C., Railsback, S.F., Robbins, A.M., Robbins, M.M., Rossmanith, E., Ruger, N., Strand, E., Souissi, S., Stillman, R.A., Vao, R., Visser, U. and DeAngelis, D.L. (2006). A standard protocol for describing individual-based and agent-based models. *Ecological Modelling*, 198(1–2), 115–126.

Grignard, A., Taillandier, P., Gaudou, B., Vo, D.A., Huynh, N.Q. and Drogoul, A. (2013). *GAMA 1.6: Advancing the Art of Complex Agent-Based Modeling and Simulation, International Conference on Principles and Practice of Multi-Agent Systems, PRIMA 2013: PRIMA 2013: Principles and Practice of Multi-Agent Systems*, 117–131.

Helms, A.C. (2003). Understanding gentrification: an empirical analysis of the determinants of urban housing renovation. *Journal of Urban Economics*, 54, 474–498.

Leao, S.Z., Troy, L., Lieske, S.N., Randolph, B. and Pettit, C. (2017). A GIS based planning support system for assessing financial feasibility of urban redevelopment, *GeoJournal*, 1–20.

Levy, S., Martens, K., van der Heijden, R. and Filatova, T. (2013). Negotiated heights: an agent-based model of density in residential patterns. *Proceedings of the 2013 CUPUM, Computers in Urban Planning and Urban Management*, 2–5 July 2013, Utrecht, The Netherlands.

OECD (2012). *Compact City Policies: A Comparative Assessment*. Paris: OECD.

Parker, D.C. and Filatova, T. (2008). A conceptual design for a bilateral agent-based land market with heterogeneous economic agents. *Computer, Environment and Urban Systems*, 32(6), 454–463.

Pettit, C. and Pullar, D. (2004). A way forward for land use planning to achieve policy goals using spatial modeling scenarios. *Environment and Planning B: Planning and Design*, 31, 213–233.

Pinnegar, S., Randolph, B. and Freestone, R. (2015). Incremental urbanism: characteristics and implications of residential redevelopment through owner-driven demolition and rebuilding. *Town Planning Review*, 86(3), 279-301.

Polhill, J.G., Parker, D., Brown, D. and Grimm, V. (2008). Using the ODD Protocol for describing three agent-based social simulation models of land use change. *Journal of Artificial Societies and Social Simulation*, 11(2), 3.

Randolph, B. (2006). Delivering the compact city in Australia: current trends and future implications. *Urban Policy and Research*, 24(4), 473–490.

Rawlinsons (2014). *Australian Construction Handbook*, Rawlinsons Publishing.

Schelling, T.C., (1978). *Micromotives and Macrobehavior*. Cambridge, MA: Harvard University Press.

Taillandier, P., Vo, D.A., Amouroux, E. and Drogoul, A. (2010). GAMA: A simulation platform that integrates geographical information data, agent-based modeling and multi-scale control. International Conference on Principles and Practice of Multi-Agent Systems, PRIMA 2010. Principles and Practice of Multi-Agent Systems, 242–258.

Troy, L., Easthope, H., Randolph, B. and Pinnegar S. (2015a). Renewing the Compact City: Interim Report. City Futures Research Centre, UNSW, Sydney.

Troy, L., Easthope, H., Randolph, B. and Pinnegar, S. (2017). It depends what you mean by the term rights: strata termination and housing rights. *Housing Studies*, 32(1), 1–16.

Troy, L., Randolph, B., Pinnegar, S. and Easthope, H. (2015b). Planning the end of the compact city? In Proceedings of the State of Australian Cities Conference 2015, 9–11 December 2015, Gold Coast, QLD, Australia.

UN Habitat (2013). *Planning and Design for Sustainable Urban Mobility: Global Report on Human Settlements.* New York: United Nations.

8

ARCHITECTURE, GIS AND MAPPING

Peter Charles and Richard Reed

Introduction

This chapter uses a conceptual framework based on three approaches which can assist to examine the growth of and change in urban systems in contemporary cities based on 'observation', 'simulation' and 'speculation'. The main objective is to examine to what extent GIS and mapping can be effectively applied in the architecture discipline and the property environment. In addition, the research investigates if these approaches provide an adequate framework for operating in a projective capacity and to what extent they could be woven together to provide a more robust model relevant to architecture. The underlying argument is based on the premise that the nature of the contemporary city has shifted towards a fragmented and dispersed state due to a variety of forces. To date it has yet to be determined to what extent other methodologies such as GIS and mapping can be utilised to better understand these complexities, which therefore forms the basis for this chapter.

The foundation for this chapter is based on the premise that most cities are being perceived as slow moving entities. Therefore, a study of the changes that occur over time requires different viewpoints to deal with varying spans of time. These viewpoints also need breadth and depth to deal with associated complexities of non-spatial systems in other realms such as economy, law and politics, as well as the ability to zoom both in and out. This is where the application of GIS and mapping are ideally positioned to contribute. In this chapter there are case study examples to examine the practical implications of the application of GIS in an urban environment such as a central business district. For the purposes of this chapter the research question is: *To what extent can GIS and mapping enhance the application of architecture in an urban context?*

Urban designers use GIS and mapping in various forms to study layers of spatial data, while planners also use mapping to assess effects of legislative planning restrictions. However, in direct contrast, there is relatively limited use of GIS and mapping in the architecture discipline. Therefore, to address this shortfall, architecture needs to develop responsive and flexible spatial models focused on the qualitative effects of spatial morphology changes over time. The starting point for an analysis into architecture and GIS is to consider the three main

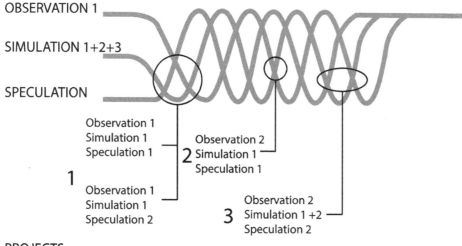

OBSERVATION 1

SIMULATION 1+2+3

SPECULATION

Observation 1
Simulation 1
Speculation 1

Observation 2
2 Simulation 1
Speculation 1

1

Observation 1
Simulation 1
Speculation 2

Observation 2
3 Simulation 1 +2
Speculation 2

PROJECTS

FIGURE 8.1 Interwoven methodology of three strands
Source: author.

approaches as shown in Figure 8.1: (i) *'observation'*, i.e. referring to the existing city in both the past and present, (ii) *'simulation'*, i.e. proposing alternative futures and (iii) *'speculation'*, i.e. presenting options of possible paths into the future. Over time these three approaches become interwoven and inter-related with each other. The balance of this chapter is based on these three approaches. Note that Figure 8.1 also includes three examples of different projects which are undertaken based on the sequence of an observation, a simulation and then speculation.

The conceptual model in Figure 8.2 then emphasises how the three approaches can be applied in the built environment; however, this diagram is a reflective rather than a guiding model. In other words, it is specific to the working method employed in this particular aspect of the project, although importantly is not how it might always be. This distinction between theory and practice is critical here, as the crossover is much more complex in actual practice and the process of discovery of the model occurs in a simultaneous manner. The very nature of design research is such that the initial model changes as a project evolves and simultaneously affects, and is affected by, the model's framework in a feedback loop.

Literature review

This section reviews previous studies of the observation, simulation and speculative approaches within the context of architecture, the built environment and mapping.

Observation approach

From an architectural perspective, an observational model ensures the city is viewed as the combined production of complex processes of cultural and material production, rather than simply as a collection of physical artefacts. This process is the consequence of human behaviours set into patterns and systems mirrored through a variety of instruments such as

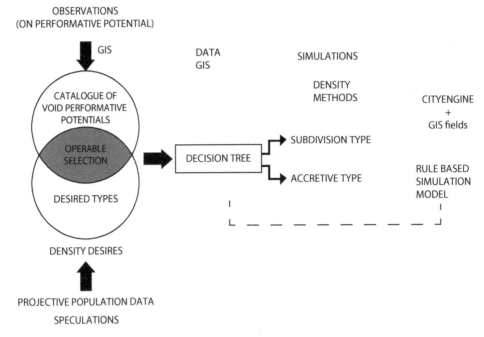

FIGURE 8.2 Conceptual diagram
Source: author.

economies and legislation. This is in contrast to a common *'built environment'* perspective of a city, which focuses predominantly on the physical structure of a city. Nevertheless, this architectural perspective and the complex forces acting on any space have previously been examined for an urban structure perspective; for example, it was argued that these forces produce a unique character and are inseparable from their atmosphere (Lefebvre 1994). The foundations of the observational method is based on reframings of the city through changing perceptions and experiences in the city, rather than being purely analytical. An early example of this reframing is a reorganised map of Paris in the wake of the 1968 student uprisings which was produced by wandering aimlessly through the streets on *'dérives* which is translated as urban drifts'* (World Policy Journal 2016). With this approach, there were certain areas in Paris recomposed in relation to the psycho-geographic experience of the individual observers rather than just on the physical connection. A decade later, Venturi *et al.* (1977) published *Learning from Las Vegas*. Although the work was not radical in intent, it repositioned how urbanism was viewed; for example, the Las Vegas Strip is organised primarily geared towards the automobile in a similar manner as for many cities in the US.

The benefits of the observation method include the availability of detailed information about the urban environment. Although observation is the only method to study relationships that have evolved, not being planned, it is accepted that a critical distance must be maintained to ensure it is effective. In addition to being a method, observation embraces a holistic outlook and develops a critical sensibility that has been referred to as critical realism (Ewing 2011). Overall the three major critiques of the observational method can be summarised as follows:

(i) When the everyday subject shifts into an 'exotic other' as a form of entertainment with no critical distance;
(ii) When the subjective bias of the observers affects the lens of research (note: this is also a strength); and
(iii) The potential for a designer to be complacent in considering historical solutions remain valid.

The most successful use of the observational method is with a combined approach with speculative and forecasting practices, but only when the boundaries between each are clearly defined and the research methods are applied projects or conceptual problems.

Simulation approach

The first key development towards the structuring of procedural simulation of problems and alternatives in relation to architecture was observed in 1964 (Alexander 1964). This encouraged the use of conceptual models by different practitioners which differed in structure and approach whilst serving as benchmarks for comparison. However, a key challenge when solving multiple conflicting interests of landowner stakeholders for the optimal negotiated path of a highway through an urban complex area provided a practical site for their explication. The relatively low uptake of urban behaviour simulations could arguably be due to the disciplinary divides between architecture, engineering and planning; note that these models have largely originated in engineering and planning.

The second key development was the use of 'agent-based models' (ABM) in 1986, which successfully simulated the behaviour of flocks of birds; however, this was not undertaken by modelling the entire system but focused on the behaviour of individuals with only a local level awareness of their neighbours. In turn, this understanding fed into ideas about emergence, collectives and self-organisation. Although agents were restricted to the definition of key lines of traffic flow networks, the first substantial exhibition of an urban ABM in the commercial arena was held in 2006 – for example refer to 'The Kar tal Pendik' (Hadid 2017). Since then some new models have emerged and articulated the potential translations of generative and procedural models into thinking about the city (Batty 2013). Arguably the current void in the literature, however, is that there is no grounded study of emergent urban behaviour which also takes physical constraints of land ownership restrictions or ascribes any performative role to voids.

An observed earlier trend was the use of computer-aided design (CAD) theory, which stemmed from research at the MIT Architecture and Planning Department in 1977 (Mitchell 1977). With this approach the focus shifted away from an instrumental approach through linear decision-making systems towards exploring the innate nature and structure of generative processes that digital procedures offer. However, this would not be referred to an explicit interest in local-level analysis as the bottom-up approach, as it was in its infancy. More recently, the amount of algorithmic work in architecture has increased following these early developments. For example, *Objective Optimization (MOO)* evolved from the idea of Genetic Algorithms (GA) introduced by John Holland in the 1970s, and was first seriously implemented by John Frazer in the early 1990s since it feeds into a theory of performative criteria; a relevant contemporary example of the extension of the field was published in the proceedings of SIMAUD 2015 titled 'Optimizing Creatively in Multi-Objective Optimization' by Ashour *et al.* (2015).

The two most useful aspects of simulation are the ability to simulate time on a step-by-step basis and the ability to explore multiple possibilities; both are crucial in relation to the overall goal of the research being to explore new urban models looking at change and growth in time. The immediate goal of the simulation of urban behaviour may be to look at multiple possible scenarios simultaneously, but this is not its purpose.

> … computers can supply the means to explore these other possibility spaces in a rigorous way because the interactions in which capacities are exercised can be staged in a simulation and varied in multiple ways until the singular features of the possibility space are made visible.
>
> (DeLanda 2011, p.20).

Note that the reference to '*possibility space*' was not referring to the possible simulated scenarios, but to a topological diagram which outlines the structure and relationships between possible paths that the scenario could assume. It is the discovery of the nature of the problem itself in the process that is important. To return to the distinction between the consolidative model and the exploratory model, the purpose of the exploratory model is not the outputs, rather the exposure of recurring patterns that start to inform specific debates about an urban system rather than just generalities. Arguably the underlying historical parallel between the digital computer and simulation makes the purpose of simulating the city somewhat nebulous; however the city has always provided the perfect subject of a complex system or ecology for which to test cognitive abilities upon. It is in this simultaneous role of subjective model and representation that simulation is located, being both a mirror and a self-perpetuating series of objectives.

> From the time when digital computers were invented over half a century ago, the idea of simulating large-scale, extensive systems such as cities became prominent. Computer models of their form and structure as patterns of location and of interactions or flows providing the glue that binds various economic, social, and land use activities were first proposed in the 1950s. The kinds of theory invoked then, on which the design of such models was predicated, was and still largely is based on the social physics of movement and potential …
>
> (Batty 2013, p.271).

It is important to note that undertaking a simulation has been primarily based on the physical sciences and not the biological sciences; in contrast this study is concerned with growth, decay and change, which are predominantly the domain of biological sciences such as biology and ecology. Referring to Banke's (1993) distinction between 'exploratory' and 'consolidative' models of simulation, Batty (2013) provided a useful point for departure for understanding the purpose of simulation models as follows:

> the consolidative style tends to focus on the construction of models that might ultimately provide accurate or focused predictions in contrast to exploratory models that will never do so, but are used to define salient characteristics and to "inform" the debate over particular problems
>
> *(Batty 2013, p.273).*

From an architectural perspective the process of undertaking simulation is productive for the following reasons:

(a) Identifying the nature of the structure of a whole system, and its processes that illustrate the relation of its parts sorted into variables and events;
(b) Recording the scale and nature of change;
(c) Breaking down complexity into smaller pieces to study;
(d) Identifying reoccurring patterns and emergent phenomena;
(e) Looking at processes in a step-by-step iterative manner in a 'freeze frame' approach; and
(f) Tracing patterns of stability that reoccur over multiple possible scenarios and speculating on their causes.

A simulation is very effective when examining long-term growth and changes in a system. Furthermore, the subsequent systemic and localised patterns of effect that are disconnected events in space or time cannot be noticed when restricted to street level observations. The simulation of urban environments may offer new perspectives on overall systems, but a simulation does not offer any perspective on the spatial qualities as experienced from human perspective either collectively or individually. On the other hand a negative effect related to this problem of exclusion is over-abstraction. Digital simulations require inputs and sets of rules that govern the interactions between these inputs in explicit clean and closed systems, although this can lead to a tendency to make them over-simplifications. In turn it cannot replicate the all-important unpredictable and unwanted interactions which occur in a city.

Speculative approach

Forecasting or speculative practices are largely fictional and relate to the future, attempting to illuminate a possible future option. What typifies a speculative practice is presuming a set of *'what if'* assumptions to be true, based on fictional parameters for a potential future scenario. With regards to modelling growth and changes in a city, reference is traditionally made to different fictional parameters regarding population and growth over cultural, political and aesthetic parameters.

The two primary types of speculative practice are:

(a) Speculative scenarios: the amplification of existing phenomena and conditions by assuming certain future *'what if'* conditions to be true and operating on this basis through fictive scenarios.
(b) Speculative growth: speculating on a parameter such as population growth or decline, under certain conditions, and then deploying a growth model to accommodate for that growth or decline scenario.

The focus in this section is concerned with the image of the city as it relates to growth and change, then the changing nature of these practices in direct response to the evolving nature of the various cities. Early examples include projects such as Claude-Nicolas Ledoux's design for the Saltworks at Arc-et-Senans in 1773; Ebenezer Howard's Garden City of 1898; Le Corbusier's Plan Voisin of 1925; Frank Lloyd Wright's Broadacres City of 1932. These works confirm that urban speculations are not new phenomena but they specifically focus on contemporary scenarios related to change, growth and decline in an existing metropolitan city within the climate of contemporary organisational shifts. These shifts can be summarised as transitioning from the era of peak energy and automobiles to a remodelling of the city as a performative ecological system, coupled with the fragmentation and dispersal of urban form due to these organisational forces. The most relevant projects related to this topic of growth are listed as:

- Metabolism;
- Landscape Urbanism (Shane 2011); and
- some key later projects by Team 10 and early works of OMA (Gargiani 2007).

The projects of Team 10 in the 1960s, as collected in the essay 'Identifying Mat Building' (Smithson 1974) in 1968, clearly showed a common strategy for flexible urban growth, distancing themselves from the city as machine model of early modernists and the Congrès Internationaux d'Architecture Moderne (CIAM) into a system based on an ecological growth model. The Free University exemplified this approach and was based on scalar modular units that fit into an expansive horizontal mat which could accommodate change; however it seemed to struggle to move beyond horizontality and ownership concerns.

Throughout the 1960s and 1970s the Metabolists continued this trajectory of growth and replaceability, where they successfully transferred the rhetoric from a machinic model to a biological model of growth and change. Through deployment of their vertical vocabulary, such as joint cores, they were better equipped to speculate on expanding into the sea and sky to accommodate population growth. Whilst the Metabolists focused on upwards growth in the sky, the Landscape Urbanism movement of the 1990s speculated on possible futures through a model that transitioned from decline of industry to reestablishing ecology on the ground plane. It should be noted that the Metabolists and Landscape Urbanists shared the staging of systems in time to facilitate urban growth and decline. These staged systems of decline were exhibited in the unsuccessful 1999 Downsview Park Competition by Field Operations and Nina Marie Lister (Holling *et al.* 2014), where the layers allowed flexibility to accommodate uncertainty; it was accompanied by a new ecological agenda focused on urban renewal through speculations on decline not growth, reframed through landscape rather than buildings.

The primary use of speculation as such is to speculate on future outcomes contingent on the amplification of certain parameters and conditions, then to set down a desirable outcome and create a road-map on how to achieve them through means available today. A useful speculative methodology highlights conditions in the present day and offers up key points of action on how to use these conditions to coerce the transition towards the desired future. Furthermore, this method may use both utopian and dystopian frameworks to highlight key issues that need to be addressed. A less useful speculation offers up a vision of a future scenario in its complete state with no transitional steps to achieve the objective with regards to existing physical or cultural conditions. A common criticism of speculative projects is they are unrealistic; however, achieving the realisation is not usually the purpose of a speculative project. Arguably the key purpose of a speculative project is often to galvanise or coerce opinion through the provision of alternate projections that comment on current conditions. However a common challenge with any large urban speculative proposition is often the requirement of mobilising capital and decision-making centrally, being a practical impossibility in a contemporary democratic environment. A good speculation is aware of these limitations and is not constrained by them, but it makes suggestions for their resolution in non-architectural spheres. In contrast a bad speculative model allows either the realities to entirely restrict the project or totally ignores them.

The key difference between the Metabolists' speculative models and the majority of other models is the support centrally from policy, legislation and finance in both the private and public spheres. With this speculative model of growth and decline, the use of the void is arguably because the model has shifted away from a model that understands speculation as a pure formal growth model, but is a model that integrates disparate and uncertain political, economic and cultural influences.

Research question

For the purposes of this chapter the research question is: *To what extent can GIS and mapping enhance the application of architecture in an urban context?*
This question will be answered via the use of multiple case studies using GIS.

Research methods and analysis

This section discusses the research methods/analysis and is structured on the observation, simulation and speculation approaches.

Observation methodology applied to voids

The emphasis has been placed on the identification of voids through observation and applying them to the subject of the voids. The approach involves assembling and curating drawings and photographs and notes for comparison that explore the qualitative aspects of voids in the present, as well as their transitional nature in time. There are three observational types in existence: (i) historical formation and growth, (ii) existing use and (iii) potential future use. Accordingly, GIS acts as a cataloguing system where relevant fields can be entered. This approach intends to define the qualitative aspects of voids which is achieved through measured quantitative means by extracting aspects of each via layered drawings that curate and separate them through the use of GIS. For example, the sites in different voids in a central city centre perform roles across at least two program types and exhibit an evolved level of 'civicness'. Therefore the intent is to identify what their common qualities actually rely on. This analysis examined the levels of porosity to external areas, adjacency and access to surrounding programs, physical boundaries compared to ownership boundaries, visual access, transitions from previous uses to existing uses in time, visible and invisible infrastructures, activity patterns and their geometric qualities. It is anticipated that in most cases the voids that transfer in use were previously used as car parks.

Visible infrastructures such as bollards, fences and cameras play a large role in the control and surveillance of these civic voids, often by individuals. The general intent behind these studies has been to provide a feedback mechanism and also material for projective use in the simulation of larger aggregations of these voids. Reference to Figure 8.3 highlights the use of mapping in identifying voids in a high-density urban area, with Fitzroy being a residential suburb located in inner-city Melbourne.

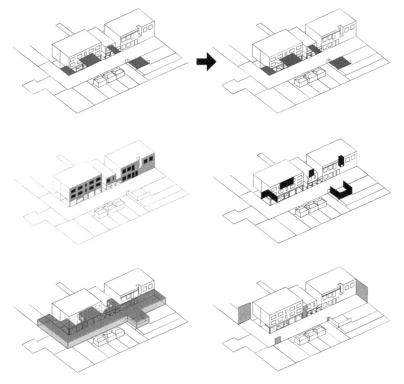

FIGURE 8.3 Elbow void: an urban void consisting of a spine of a public road and an accumulation of private car parks
Source: author.

Case study A – void analysis in residential areas

The voids highlighted in Figures 8.4 and 8.5 illustrate varying topologies for different porous spaces examined through a method of mapping out their individual uses. It is possible to draw out individual qualities of their actual use, which has evolved over time on a different trajectory from the original intention.

Case study B – void analysis in an industrial area

This case study applied a similar methodology of void assessment to an industrial site located in Fishermans Bend, Melbourne, based on a block-by-block analysis. As shown in Figure 8.6, the site was bounded by Salmon Street, Williamstown Road, Plummer Street and a park; in turn this provided a highly connected address but with a noticeably smaller existing grain than most other sites. This observation is concerned with previous formation processes and patterns of future probability of voids arising to determine where the most appropriate place of action is. The objective was to extract these observations to simulate the probability of certain parcels becoming useful and to link parcels with future public roles for use at a strategic level.

The next step was to catalogue the existing voids on sites within the Fishermans Bend project area by comparing and contrasting voids based on their common characteristics. The existing voids in Fishermans Bend and their qualities were compared in a matrix as shown

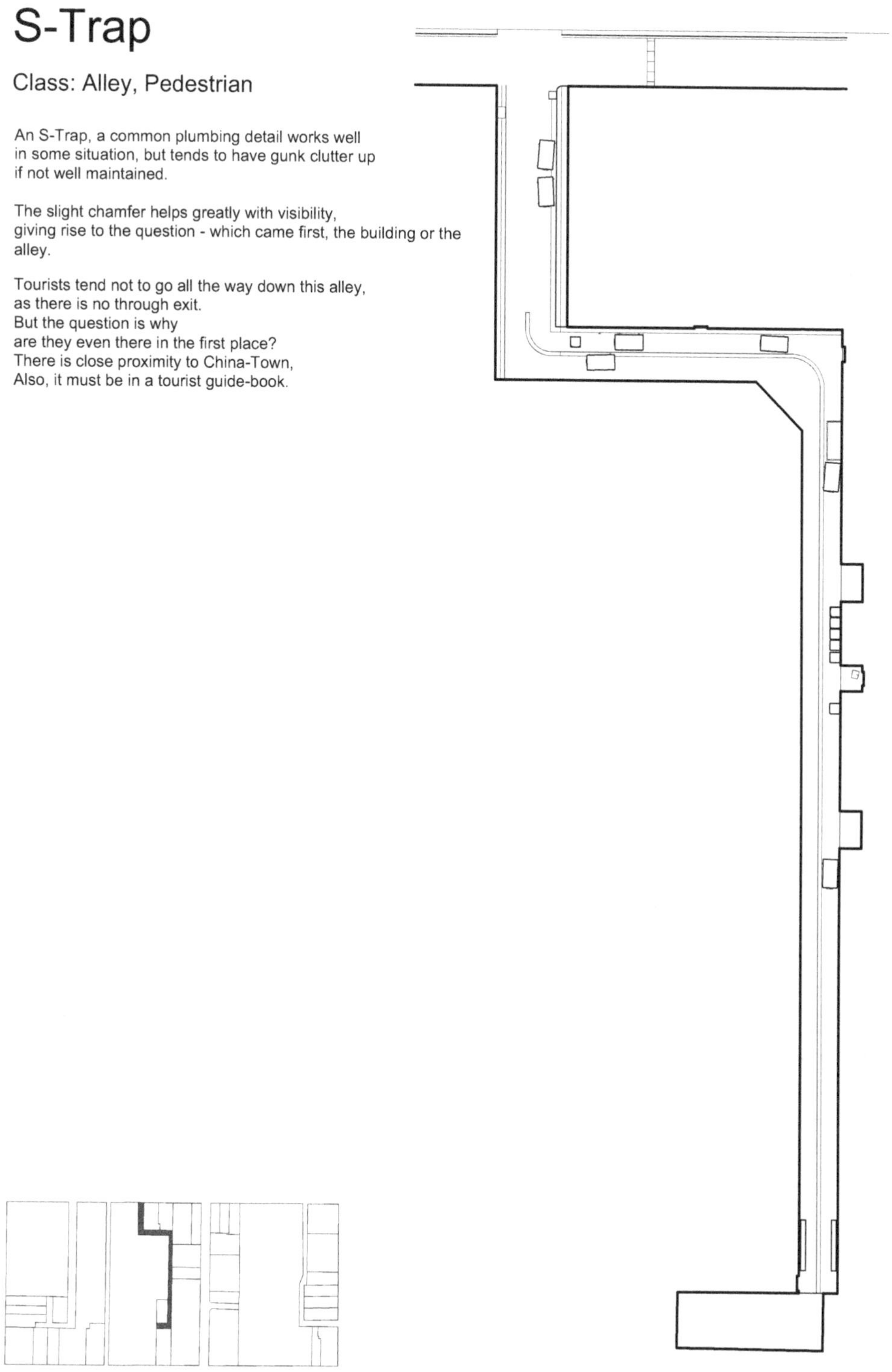

S-Trap

Class: Alley, Pedestrian

An S-Trap, a common plumbing detail works well
in some situation, but tends to have gunk clutter up
if not well maintained.

The slight chamfer helps greatly with visibility,
giving rise to the question - which came first, the building or the
alley.

Tourists tend not to go all the way down this alley,
as there is no through exit.
But the question is why
are they even there in the first place?
There is close proximity to China-Town,
Also, it must be in a tourist guide-book.

FIGURE 8.4 S-trap void – pedestrian area
Source: author.

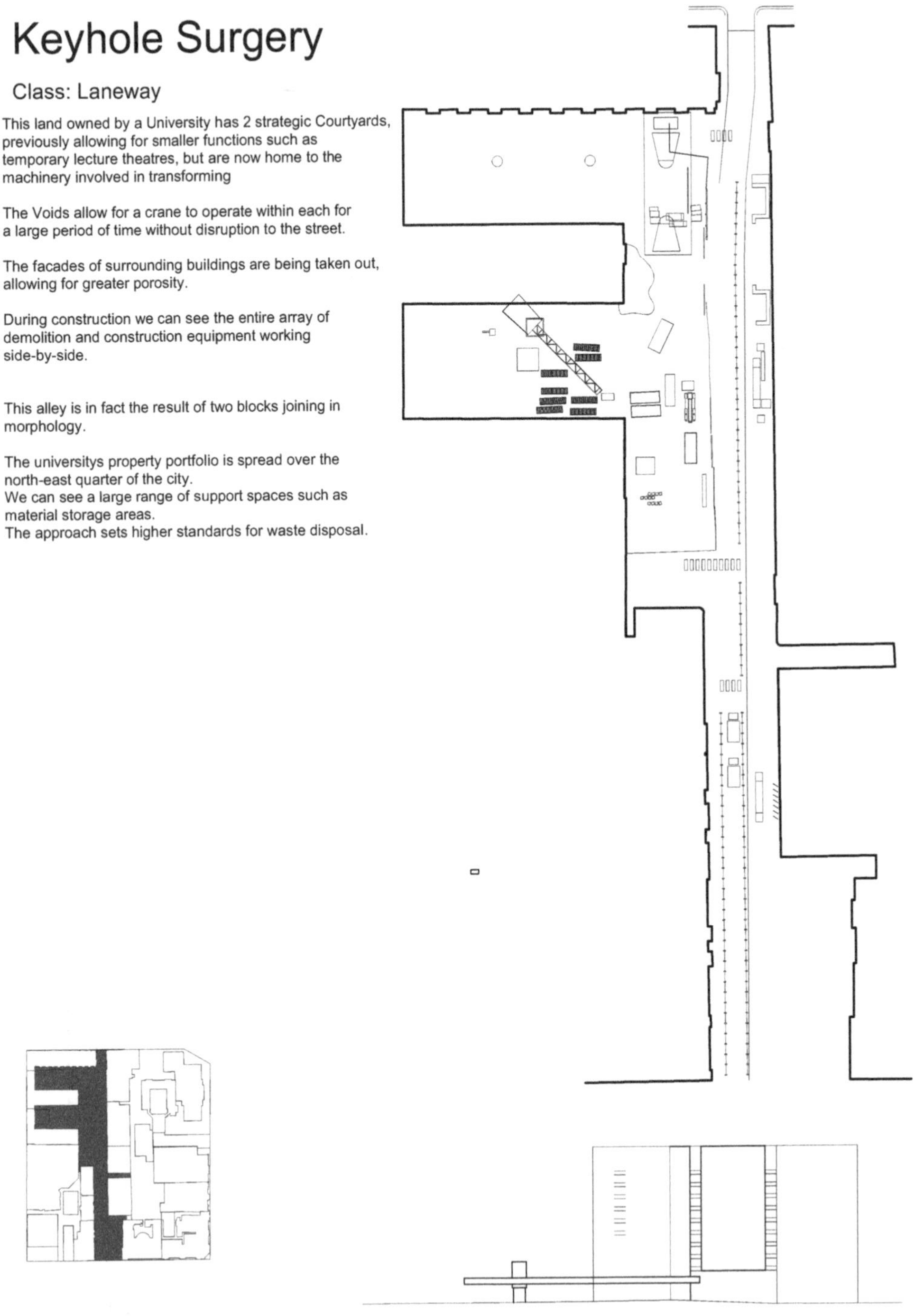

Keyhole Surgery

Class: Laneway

This land owned by a University has 2 strategic Courtyards, previously allowing for smaller functions such as temporary lecture theatres, but are now home to the machinery involved in transforming

The Voids allow for a crane to operate within each for a large period of time without disruption to the street.

The facades of surrounding buildings are being taken out, allowing for greater porosity.

During construction we can see the entire array of demolition and construction equipment working side-by-side.

This alley is in fact the result of two blocks joining in morphology.

The universitys property portfolio is spread over the north-east quarter of the city.
We can see a large range of support spaces such as material storage areas.
The approach sets higher standards for waste disposal.

FIGURE 8.5 Keyhole surgery void – pedestrian area
Source: author.

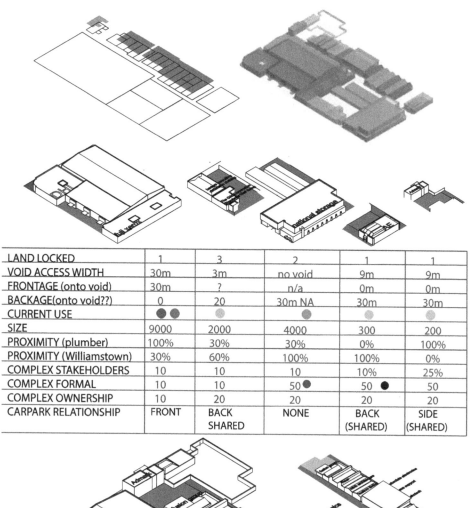

LAND LOCKED	1	3	2	1	1
VOID ACCESS WIDTH	30m	3m	no void	9m	9m
FRONTAGE (onto void)	30m	?	n/a	0m	0m
BACKAGE(onto void??)	0	20	30m NA	30m	30m
CURRENT USE	●●	●	●	●	●
SIZE	9000	2000	4000	300	200
PROXIMITY (plumber)	100%	30%	30%	0%	100%
PROXIMITY (Williamstown)	30%	60%	100%	100%	0%
COMPLEX STAKEHOLDERS	10	10	10	10%	25%
COMPLEX FORMAL	10	10	50 ●	50 ●	50
COMPLEX OWNERSHIP	10	20	20	20	20
CARPARK RELATIONSHIP	FRONT	BACK SHARED	NONE	BACK (SHARED)	SIDE (SHARED)

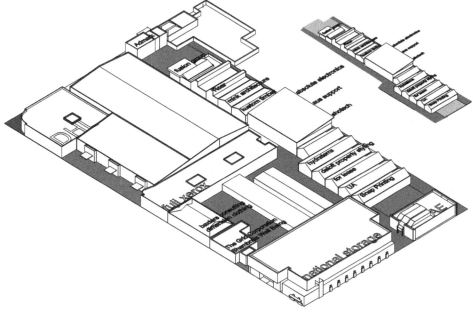

FIGURE 8.6 Case study of the voids in Fishermans Bend, Melbourne
Source: author.

in Figure 8.6. The process of comparing and contrasting their similarities and difference highlighted three major categories of their potential role in the future; this approach also produced an image of the area and built a coherent map of disparate parts with a shared history. The voids are often not related to complexities limiting development due to physical constraints, but are linked to a disconnect of spatial opportunity and the line of business of the occupant. The study was undertaken on an individual block-by-block basis and has not been as insightful as anticipated, with the results suggesting the blocks are more generic than expected. Furthermore, most voids are actually expansive car parks associated with buildings or are entire parcels that have a high chance of improvements being demolished with potential new development. The majority of voids actually remain as voids outside parcels; for example as kerb space.

Simulation methodology applied to voids

This section investigates the exploration of voids via a simulation process by drawing from existing methodologies (e.g. the work of 'Team X Mat Building') and applying them to the subject of the voids. The simulations are comprised of a series of small components which are intended to work collectively as one model.

Emergent parcel kit

The intent of the overall simulation section of the model is to have a series of components that act collectively in sequence, assessing both individual parcels in relation to the porous spaces located on them and also their existing density. The objective is to address the complexity of city plots rather than consider all as equal, with the model harnessing this differentiation of plots to produce different effects. This will produce varying results that will vary between a simulated reality and a desirable outcome.

The model comprised of varying toolkit components, rules and time iterations that worked in relation to each other as follows:

(a) GIS decision-making tree based on intersections;
(b) Rules based formal density simulation methods based on subdivision and accretion;
(c) Speculative scenario variables that act as forces affecting the decision tree; and
(d) Time iteration.

The decision-making tree is informed by the data overlays as shown in Figure 8.7. With this approach, the variations become evident via the intersection of data in GIS, in turn highlighting which density simulation models should be employed.

The diagrams in Figures 8.7 and 8.8 and associated fields in a table format highlight how the same porous spaces can be allocated various different values for different attributes and potential functions. These semantic relations are relatively new to the field of architecture, which typically only works with geometry. These major four fields include vertical, farmlane, park rank and forest rank. Importantly all descriptions require a complex analysis regarding the height of surrounding buildings, orientation of the lanes and width of lanes to determine varying scenarios for these different programs. This is potentially quite an important and encouraging step in urban design when considering performative concerns to porous

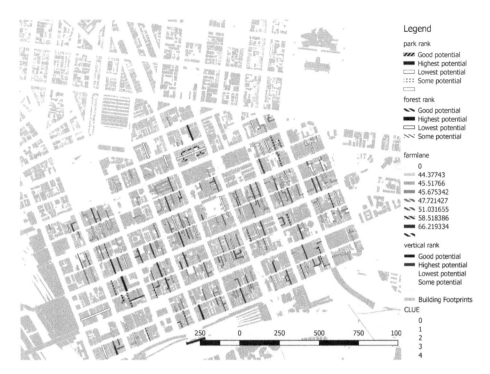

Legend

park rank
- ▨ Good potential
- ▬ Highest potential
- ▭ Lowest potential
- ⠿ Some potential
- ▭

forest rank
- ➘ Good potential
- ▬ Highest potential
- ▭ Lowest potential
- ⋈ Some potential

farmlane
- 0
- 44.37743
- 45.51766
- 45.675342
- 47.721427
- 51.031655
- 58.518386
- 66.219334

vertical rank
- ▬ Good potential
- ▬ Highest potential
- Lowest potential
- Some potential

- ▨ Building Footprints
CLUE
- 0
- 1
- 2
- 3
- 4

250   0   250   500   750   100

FIGURE 8.7 Case study of laneways in Melbourne CBD, highlighting vertical surrounds
Source: author, City of Melbourne.

Legend

park rank
- ▨ Good potential
- ▬ Highest potential
- ▭ Lowest potential
- ⠿ Some potential
- ▭

forest rank
- ➘ Good potential
- ▬ Highest potential
- ▭ Lowest potential
- ⋈ Some potential

farmlane
- 0
- 44.37743
- 45.51766
- 45.675342
- 47.721427
- 51.031655
- 58.518386
- 66.219334

vertical rank
- ▬ Good potential
- ▬ Highest potential
- Lowest potential
- Some potential

- ▨ Building Footprints
CLUE
- 0
- 1
- 2
- 3
- 4

250   0   250   500   750   100

FIGURE 8.8 Laneways in Melbourne CBD, highlighting use of potential forest ranking
Source: author, City of Melbourne.

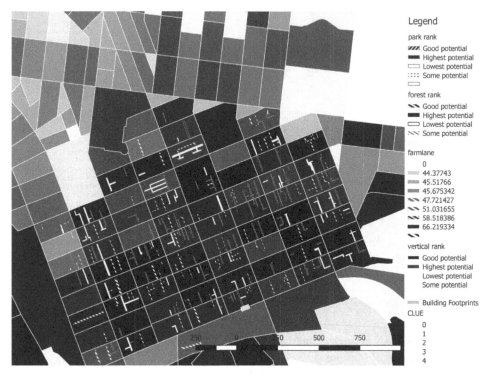

Legend

park rank

▨▨ Good potential
▬▬ Highest potential
▭ Lowest potential
∷∷ Some potential
▭

forest rank

↘↘ Good potential
▬▬ Highest potential
▭ Lowest potential
↘↘ Some potential

farmlane

0
44.37743
45.51766
45.675342
47.721427
51.031655
58.518386
66.219334

vertical rank

▬ Good potential
▬ Highest potential
Lowest potential
Some potential

▬▬ Building Footprints

CLUE

0
1
2
3
4

FIGURE 8.9 Case study of combined employment and laneway data in Melbourne CBD
Source: authors, City of Melbourne 2017b.

spaces for distinct programming in a dense urban area. Therefore full consideration of potential buildings to the empty space between them and the relationship between each have to be reconsidered as a field relationship, rather than the traditional model that only considers the program of buildings.

Figure 8.9 is based on information collected for the bi-annual *Census of Land Use and Employment* (CLUE) dataset as coordinated by the City of Melbourne (City of Melbourne 2017a); this census is focused on employment per block, overlaid with a dataset titled 'laneways with greening potential'. When GIS is used to combine the types of porous laneway spaces according to their observed environmental performances with data about existing employment density, there arises are a new set of varied relational intersections. It is then through these intersections that it is possible to see what imbalances exist and use this as a way of modulating desired differentiation within the projective model for adding a new density. For example, if a high density existed with a low rate of porous laneway space on a specific block, then it is possible to add a specific level of density through the accretion methodology. Furthermore, these values can be used to determine which blocks would benefit from porous spaces being utilised as parks to add value as economic attractors, or alternatively as productive agricultural zones for blocks with more residents than employment.

There are two simulation strategies which represent two different sub-methods of projecting spatial parameters on to a parcel of land where both work in relation to the parcel kit. The first strategy is 'projective subdivision' which is easier to control in relation to its parcel boundary; the second strategy is 'aggregation' which incorporates a more adaptive growth mechanism.

(i) Simulating voids through projective subdivision

This strategy harnesses subdivision and merging to introduce voids. It seeks to emulate the evolved nature of a mature city via fast-tracking the subdivision process of large parcels. For example, this case study refers to those parcels in Fishermans Bend, Melbourne, and is undertaken by projecting desirable voids at dispersed patterns. It establishes a series of voids as 'no build' zones which can cater for a range of possibilities; in addition it seeks a balance to not destroy the potential for larger buildings, but discourages simplistic solutions such as podium plus tower models. This projective strategy seeks to implant voids which have performative qualities beyond being left-over spaces. In this model the developers of parcels would inherit ground plane and upper floor void easements.

This approach works very well as a 'block strategy' and is useful for examining how parcels inter-relate to each other, rather than leaving voids as left-over by-products. In addition, this strategy is useful as it is fast and controllable. The voids are known at an individual scale and although there may be a 'hunch' of what they may become when they combine collectively, this aspect is uncertain until trialled. The weakness is that it scales the voids when projecting them on to existing parcels, which is not ideal; in addition, many voids have shapes that are not rectangular.

(ii) Simulating voids through projective aggregation

This exercise uses ideal voids and grows them accretively in sequential void units. This is undertaken via rules that govern small local level rules which build into a larger unknown whole. This approach is effective at dividing growth into a number of steps and therefore is useful for ensuring optimal time and timing. Some voids actually become redundant; therefore it is possible to couple performative roles of voids and the relationships that form between the combined voids to collectively produce new relationships.

This strategy works as a 'housing strategy' on entire blocks and identifies useful geometries for collective voids and also making new relationships between voids at a combined collective level which are either not present or necessary at the individual scale. In turn, this highlights the relationships between things which matter. Furthermore, this strategy is useful for predicting how growth may produce unknown effects from different seed points; for example, usually from the centre of a parcel. The approach used is an L-System, which is similar to the ESRT (City-Engine building software) component. However, whilst a City-Engine is limited to a five-step hierarchical tier, this component design enables a lot more flexibility and unknown results, being centred on the combination of voids.

Speculative methodology applied to voids

Applying the speculative *'what if'* model on-site involves proposing varying scenarios which are limited to blocks. Both the speculative scenario and the speculative population growth scenario can be examined in combination with both simulation and observation approaches.

Scenario (A) – population growth

The parameters for this scenario are based on providing accommodation on an average per hectare basis as follows: 300 residents, 100 office workers and 50 industrial workers. There is staged implementation as outlined below:

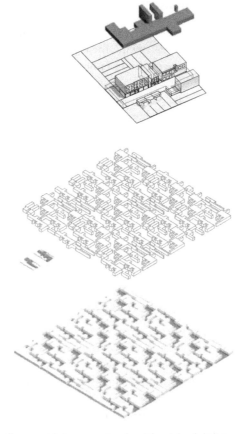

FIGURE 8.10 Projecting elbow voids in an optimal residential subdivision
Source: author.

Stage 1 transition
- 65% of total site area = unchanged parcels
- 20% of total site area = existing amplified industry
- 25% of total site area = housing and car parking (stage 2 transition)

Stage 2 transition

- 25% of total site area = unchanged parcels.

[Note: averages are per person = housing (65m^2), office (15m^2), industrial (35 m^2)].
Examples of relevant '*what if*' scenarios

What if? Existing industry was given incentive to stay and truck access must be maintained.
What if? Sea level rise of 2m occurred by 2035.
What if? Air rights became less relaxed allowing selling of air space to a neighbouring parcel.
What if? The local council could enact easements on existing land.

What if? All new dwellings must accommodate stormwater on site.

What if? Water prices increased by 500%.

What if? Fuel prices are set for a 500% increase; therefore car parking must be decommissionable.

What if? All buildings must have the same dimensions.

These scenarios below provide an insight into the project's parameters, through a refinement of the type of urbanism they create due to voids assuming a performative role.

Scenario (B) – 'What if' scenario plus growth scenario

The second case study uses a *what if?* scenario combined with a growth scenario. It also follows the lineage of the Metabolists and it assumes a rate of growth. In this applied example, the scenario asked is as follows: *What would be the scenario if the existing industry was given an incentive to stay?*

Scenario (C) – speculative scenario method

This method adopts new models of density that incorporate voids from the simulation approach through the intersection of existing densities of blocks and desired types of voids (Figure 8.11). This method follows the lineage of the Delirious New York model (Koolhaas, 1994) and employs a larger observation component than simulation, therefore setting the scene for the site by addressing the diversity and plurality of historical and potential trajectories through compiling material that covers the various identities of the site. At a territorial scale the historical legacy of the initial blocks is very strong. The desired voids in this scenario are voids that can accommodate hybrid functions of manufacturing and housing, easing in the transition from manufacturing to housing by accommodating both in the interim period. The speculative scenario that follows on from the observation and simulation approaches is outlined here.

> Observation: concrete production void
> Simulation method: project ideal subdivision
> *What If?* scenario: manufacturing incentive

The scenario is tested by amplifying a key component of the area's truck circulation into a future neighbourhood with a population growth scenario, then attempting to overlay this proposal on to a site. The response is a strategic void that evaluates options to accommodate a truck turning circle into the central organising device of the block (Figure 8.11). Observations of concrete factories in the surrounding area inform the nature of the void, converting the observation into a speculative activity. The simulation method of subdividing blocks is used to project a number of voids on to the site. A transition scenario is imagined for each, looking at the two stages. Stage 2 aims for transition of these voids into alternate activities without disruption to the area, where car parks are provided as separate buildings. The effect is that the housing turns its back to the centre towards the street (Figure 8.12). Furthermore, the void is internalised to minimise noise and attempts to create a more active outer shell, therefore requiring exterior setbacks for pedestrian walkways and resulting in a more civic frontage

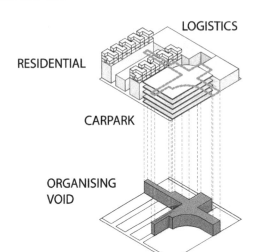

LOGISTICS

RESIDENTIAL

CARPARK

ORGANISING
VOID

FIGURE 8.11 Performative truck void
Source: author.

between housing. One of the most difficult aspects of acting speculatively is the lack of limits, parameters and purpose of a future void, and hence the relationships a void must consider as well as the form it takes. Note that for a speculative model of voids it is very important to establish performative driver for the voids, i.e. a program/activity.

The projects attempt to assign very concrete performative roles to voids such as stormwater and car parking; they also introduce metrics for measuring success, such as minimising the length of drainage areas. Note that the voids need more specificity for any successful quantification of success failure and effect. Also the setting of speculative scenarios is useful for establishing cultural frameworks. The speculative scenarios provide the basis for defining the limits and scope of a project.

Results

The following architectural scenarios were examined using GIS:

- Case study of a void analysis in a residential area. Based on examining the spine of a public road and an accumulation of private car parks (Figure 8.3).
- S-trap void – pedestrian area (Figure 8.4)
- Keyhole surgery void (Figure 8.5).
- Case study of the voids in an industrial area (Figure 8.6).
- Case study of laneways in a CBD highlighting vertical surrounds (Figure 8.7).
- Laneways in a CBD highlighting use of potential forest ranking (Figure 8.8).
- Case study of combined employment and laneway data in a CBD (Figure 8.8).
- Projecting elbow voids in an optimal residential subdivision (Figure 8.10).
- Void projected on site with anticipated growth estimates (Figure 8.12).

In each scenario the application of GIS made a valuable contribution to the observation, simulation and speculation approaches.

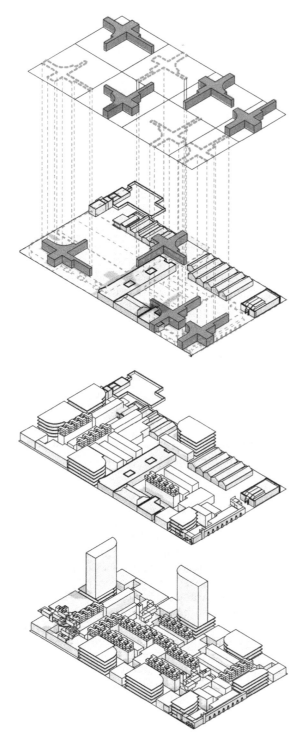

FIGURE 8.12 Void projected on site with anticipated growth estimates
Source: author.

Conclusion

With reference to the stated research question for this chapter all three approaches, namely (a) observation, (b) simulation and (c) speculation, confirmed that architecture can be enhanced by the use of GIS when considering the long-term effect on the built environment incorporating the property and real estate market. Furthermore, the three approaches highlighted different aspects of the roles that voids play over different timescales and spatial scales, hence the argument for a combined model which provides links over the different scales in order to better critique small immediate actions which can accumulate into territorial scale effects. Observation provides more tacit knowledge into the ways in which voids bind relationships between seemingly disconnected elements and provides useful qualitative frameworks for the spatial qualities of voids as experienced at a pedestrian level. Simulation proves useful at both longer periods of time, with some simulations being better at larger territorial scales and block scales. Speculation, although running as a parallel stream, relies on weaving together observations and simulations within a narrative framework which GIS can assist to provide invaluable 'what if' scenario evaluations.

Arguably the separation of the methodologies is artificial and is intended to provide a framework for exploring different combinations with clarity to distinguish between generic and ideal strategies, site restrictions and intuition; this is based on the knowledge that the projects may encompass a set of intersections. The diagram in Figure 8.1 highlights how potentially the three methodologies could examine three different intersections of the three streams, where a project may combine more than one intersection. Note that the projects are not a linear sequence of Observation → Simulation → Speculation, although there can be different weightings towards a certain strategy. The projects explore different ways of combining these methods in different orders, providing a set of potentially unique roadmaps for speculative projects. Future research should focus on applying this technique to the built environment in other cities to examine how it can be used under different circumstances, e.g. in an older established city or a new planned city.

References

Alexander, C. (1964). *Notes on the Synthesis of Form*, Harvard University Press, Cambridge MASS.

Ashour, Y. and Kolarevic, B. (2015). Find in CUMINCAD optimizing creatively in multi-objective optimization. Proceedings of the symposium on Simulation for Architecture & Urban Design, Alexandria, VA, 12–15 April 2015, 128–135. San Diego, CA: Society for Computer Simulation International.

Batty, M. (2013). *The New Science of Cities*, MIT Press, Cambridge Mass.

City of Melbourne (2017a). Laneways with greening potential. Retrieved from: http://data.gov.au/dataset/laneways-with-greening-potential. Accessed 1 March 2017.

City of Melbourne (2017b). Census of Land and Employment (CLUE). Retrieved from: www.melbourne.vic.gov.au/about-melbourne/research-and-statistics/city-economy/census-land-use-employment/Pages/clue.aspx. Accessed 1 February 2017.

DeLanda, M. (2011). *Philosophy and Simulation*, New York: Continuum.

Ewing, S. (2011) Knowing and navigating the terrain. In *Architecture and Field/Work*, S. Ewing, J.M. McGowan, C. Speed and V.C. Bernie (eds.), Routledge.

Gargiani R. (2007). *Rem Koolhaas/OMA – The Construction of Merveilles*, translated by Stephen Piccolo. London: Routledge, 118–124.

Hadid (2017). Retrieved from: www.zaha-hadid.com/masterplans/kar tal-pendik-masterplan/.

Holling, C.S. and Goldberg, M.A. (2014), Ecology and Planning. In C. Reed and N. Marie-Lister (eds.). *Projective Ecologies*, 106–125. New York, Harvard University GSD Actar Publishers.

Koolhaas, R. (1994). *Delirious New York: A Retroactive Manifesto for Manhattan*. New York: Monacelli Press.

Lefebvre, Henri (1994). *The Production of Space*. Blackwell, Oxford.

Mitchell, W. (1977). *Computer Aided Architectural Design*. New York: Van Nostrand Reinhold.

Shane, G. (2011). The emergence of landscape urbanism. In C. Waldheim (ed.), *The Landscape Urbanism Reader*. *SIMAUD*, Washington DC.

Smithson, A. (1974). How to recognise and read Mat Building. *AD*(9): 573–590.

Venturi, R., Scott-Brown, D. and Izenour, S. (1977). *Learning from Las Vegas: the Forgotten Symbolism of Architectural Form*. Cambridge, MA: MIT Press.

World Policy Journal (2016). The big question: how can governments collaborate with the private sector to provide affordable housing. *World Policy Journal*, 33(2), 1–4; doi.org/10.1215/07402775-3642440.

9

3D AND VIRTUAL REALITY FOR SUPPORTING REDEVELOPMENT ASSESSMENT

Aida E. Afrooz, Russell Lowe, Simone Zarpelon Leao and Chris Pettit

Introduction

The increasing complexity of cities demands improved planning systems. At the same time, advances in computer technologies and the advent of smart cities and big data indicate a new era for urban planning and management (Roumpani 2013). The convergence of three-dimensional (3D) modelling and virtual reality (VR) technologies provides the opportunity to build virtual cities. This in turn enables novel forms of visualisation of the built environment and potentially effective simulation of environments that planners can use to explore changes in the city before they are implemented in the 'real world' (Dodge *et al.* 1998).

3D digital models have been used in the past, mainly for the purpose of visualisation or graphical exploration of a city rather than for incorporating additional information about the city such as ontological structures, including their different attributes and interrelationships (Kolbe 2009). Nowadays the increasing demand for additional information about cities entails the necessity of advanced 3D modelling analysis and visualisation tools in a standardised fashion which can provide the opportunity to move beyond the 2D plans that are typically inserted as graphics in reports. 3D City Information Modelling (CIM) aims to achieve this and manage a range of city information data (Xu *et al.* 2014).

As noted by Klosterman and Pettit (2005), urban 3D visualisation has been identified as one of the four analytical tasks for which digital planning tools should aid planning practitioners. In the broader context of urban planning, 3D visualisation has been also used for urban disaster management, simulation of terrorist attacks and assessment of escape routes (Boguslawski and Gold 2009), air pollution modelling and monitoring (Lin *et al.* 2009), as well as vehicle navigation and emergency management (Boguslawski and Gold 2009).

Besides urban planning applications, 3D modelling can be used to support real estate and property-related activities, including the determination of land and property values. Property valuation depends on a variety of attributes, including location, shape, orientation and quality of a building (Isikdag *et al.* 2015). Although property valuation using 3D modelling is in an exploratory phase, studies show that 3D building models can enhance the process as well as the presentation and dissemination of valuation data (see Isikdag *et al.* 2015). Lately, 3D modelling and VR are more commonly being utilised for property and real estate applications, so immersive property tours are being implemented as part of the property inspection process. For instance, Ouwens Casserly is the first real estate agent in Adelaide, Australia, that is using VR walkthroughs in the property market (Millar 2015). They have partnered with the technology company TicketyView (http://ticketyview.com/) to offer prospective buyers access to properties using VR goggles.

In addition to the above-mentioned applications of 3D and VR, we consider how the immersive contextual visualisation and navigation experience of 3D/VR could additionally provide a better environment for the assessment of a development application (DA). In this context a DA is a formal request from a property owner or developer for consent by the local authority to carry out development(s) in the form of construction of new building(s) within its administrative region (NSW Department of Planning and Environment 2017). These developments include: construction of a building (residential, commercial or industrial property), making an alteration to the building, subdivision of the land, strata subdivision, and changing the building use (Randwick City Council 2017). A DA consists of a collection of documents, including application forms, site plans and consultant reports, generally presented as reports and 2D plans. This collection is then assessed by qualified professionals for the local authority against the relevant controls, such as setbacks, floor space ratio (FSR) and maximum building height. In Australia this documentation is also exhibited to the community for consultation and feedback. Many local councils are implementing DA online tracking systems as part of e-planning initiatives. However, understanding development proposals against planning controls is often not an easy task for people without expert knowledge and skills in architecture and planning. Although DA tracking enables planners and residents to track the approved DAs, such systems generally lack data visualisation tools, making it difficult for most people to envision the approved buildings, originally presented in 2D, in 3D.

This chapter extends the literature by investigating the usability of 3D modelling and VR for assessing Das, with a specific focus on Sydney, Australia. In this regard we investigate if the DA assessment process can be significantly improved by using 3D digital models and VR technology. To test this, we conduct a user experiment to investigate the impact of implementing different Level of Detail (LODs) of the virtual environment to specifically understand which level is most suitable for various tasks in the development application assessment process. Different LODs are defined by CityGML for 3D city models (Kolbe *et al.* 2005) and further described in a later section. Moreover, since VR offers a rich sensory experience for participants who can simulate and interact with 3D digital models in a controlled environment, this study also examines the capacity of incorporating VR to enhance controlled 3D virtual environments for improving the DA assessment processes in local governments.

This chapter begins by presenting the research aim and defining research questions. Then it reviews current literature related to virtual reality (VR) and provides a background for 3D and VR. The interviews with planning professionals and the experiment named 'LOD experiment' are detailed in the method section. The results from the interviews and the experiment are then reported. Finally, the chapter concludes by summarising the key contributions of this work, and we make some recommendations for more advanced analysis using 3D modelling. In addition we suggest some future research directions.

Research aims

This study consists of two aims that pertain to the practical domain of urban planning. The first aim of this research is to enhance better quality services in local government sectors by investigating the implementation of 3D modelling and VR to assist local governments prior to DA approvals and during DA assessments. The DA assessment processes at the three local councils included in this study are reviewed, and historical changes in DA assessment processes are identified. This included the use, or potential use, of 3D modelling to assist in DA assessment. This investigation has been undertaken in the effort to provide insights and recommendations regarding the implementation of 3D modelling to address local government issues regarding DA, particularly for large or complex design proposals.

The second aim of this chapter is to determine the importance of 3D modelling for better understanding urban design scenarios among professionals (i.e. urban planners and designers). Using the context of DA assessment tasks involved in a development application process, this research practically examined the effect of different levels of detail (LOD) inherent within the 3D digital models. This study aims to provide recommendations which can enable efficient use of LODs in 3D modelling for urban planning purposes. To achieve this aim, an experiment was designed using a VR headset investigating the applicability of different LODs in DA assessments.

Research questions

The following research questions are addressed in this chapter:

(i) *What are the benefits of utilising 3D modelling in local government sectors for DA assessments?* This question addresses opportunities and barriers found in practical implications of 3D modelling in urban planning and design fields, with particular focus on DA assessment processes. Focus group interviews with planning professionals are developed to address this question.

(ii) *Which LODs in a city model can assist planners in assessing DAs?* Assuming that 3D modelling can ease and enhance DA assessments, this question is asking which LODs can best meet the requirements of planners in assessing DAs. Although LOD is reasonably defined among scholars, there is no experiment-based evidence to indicate the suitability of each LOD to the assessment of varied aspects of a design scenario. Using a DA assessment process as case study in selected councils in Sydney, an LOD experiment is developed in this study to respond to this question.

Background literature review

This section presents a review of previous literature. First it describes the progress of 3D modelling in urban planning and real estate, followed by a description of the influence of different levels of details (LOD) of 3D digital models on urban planning applications. Third, technological developments of virtual reality, which added an immersive environment to urban 3D models, are discussed.

3D modelling in urban planning and real estate

In the past 3D digital models were created separately and then imported into the corresponding software. Such elemental models only showed a simple building volume extruded from a footprint, which did not allow the integration of GIS data and the conjoint analysis using terrain and surrounding urban areas (Luo *et al.* 2015). After the 3D models were built they were exported to other software to conduct additional simulations and analysis, such as wind and shadow analysis or flood and thermal simulations. Moreover, in the past many planners and geographers were not able to display theories of urban modelling using GIS (Birkin *et al.* 1996). However, nowadays such visualisations and analysis are possible using software such as ESRF's CityEngine (CE); this can be used, for example, to visualise the von Thünen model – a simple model of land use that showed how land in different locations would be used as determined by market forces (see Roumpani 2013).

Currently there are a number of research groups around the globe that are developing tools for generating large scale 3D modelling. Examples of these groups include the Urban Simulation team at UCL's Department of Architecture and Planning, the Redland redevelopment projects done by ESRI, which demonstrate the applicability of CE in urban planning, and the CityGML which was developed by Geodata Infrastructure North-Rhine Westphalia, Germany (GDI NRW). Terrestrial Laser Scanning (TLS) is another relatively new technology that is being increasingly used for creating 3D models in different fields, including architecture (see Valero *et al.* 2015), as it generates mesh and CAD models from collected 3D point clouds in a rapid and accurate way.

In planning the use of GIS tools such as CE, attempts to combine spatial analysis from ESRF's ArcMap with the 3D presentations using CGA shape grammar (i.e. script language like Python used by CE interface) to conduct a wide range of procedural modelling. Generating 3D digital models using CE has two main applications: first, spatial analysis can be used at both urban macro (e.g. regional planning) and micro levels (e.g. urban design); and second, simulation analysis combines geographical (i.e. GIS) and physical environments (Luo *et al.* 2015).

In recent years the use of 3D modelling has had to meet different requirements from the clients, which do not merely respond to visualisation purposes but also help to determine the attributes of the features in the model (Xu *et al.* 2014). City models are inherently complicated because they include both static and dynamic objects as well as companies, organisations and transportation systems (Xu *et al.* 2014); therefore a large amount of information needs to be generated. This makes the development of 3D city model imperative to effectively extract useful data from massive information. City Information Modelling (CIM) is one useful method to organise this large amount of urban information (Xu *et al.* 2014). Inspired from Building Information Model (BIM), CIM is a highly efficient and multifunctional management system developed to concentrate a wide range of information from a digital city, including building, infrastructure and property information among many more aspects (Xu *et al.* 2014). Figure 9.1 illustrates different modules of CIM.

BIM is defined as:

> the information management process throughout the lifecycle of a building (from conception to demolition) which mainly focuses on enabling and facilitating the integrated way of project flow and delivery, by the collaborative use of semantically rich 3D digital building models in all stages of the project and building lifecycle
>
> *(Underwood and Isikdag, p.86).*

Therefore BIMs contain geometric/semantic information related to buildings which varies between different LODs (Isikdag *et al.* 2015). The evolution of BIM was fast; the first evolution

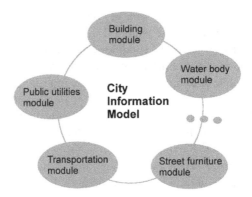

FIGURE 9.1 City Information Model (CIM) (derived from Xu *et al.* 2014)

of BIM was a shared house of information which was evolved to an information management strategy and to a construction management method (Isikdag 2015).

3D modelling can be utilised as a guide for building new districts, as it can help planners to visualise different scenarios for site selection and redevelopment –e.g. see Redlands redevelopment case study (ESRI 2014). CE 3D modelling can support planners to improve the links of both transport and infrastructure between different parts of the city – i.e. old and new, or suburban areas (Luo *et al.* 2015). The ability of generating the streets automatically allows planners to compare different street networks (Luo *et al.* 2015).

Additionally CE 3D modelling has several applications for micro scale planning, particularly for urban designers who can experience the space by virtually walking in different simulations. In the initial phase of design or planning, CE enables planners and designers to adjust the height, texture and amount of greenery, floor space ratio (FSR), skyline, building layout and open spaces (Luo *et al.* 2015). CE can also promote public participation by sharing the design through web-based platforms such as ESRF's 'ArcGIS online' or 'WebScene'. People can even check shadows by changing the time of the day, leave comments, and search locations using WebScene.

Level of detail (LOD)

Level of detail is a concept used in urban 3D digital models. In practice it connects the model to its real-world counterpart (Biljecki *et al.* 2014). Although LOD has been widely used in different disciplines such as architecture, cartography, GIS and planning, it is not a standardised concept (Biljecki *et al.* 2016). The definition of LOD varies between different practitioners in the field of architecture and urban design. For example, Biljecki *et al.* (2014) defined ten discrete LOD categories that are suitable for architectural design purposes. However, for the purposes of this study which focuses more broadly on urban planning, the CityGML definition of LOD is selected. Figure 9.2 illustrates different LODs of a 3D urban representation in virtual environment according to the CityGML classification.

CityGML defines five different levels of detail for 3D modelling (Biljecki *et al.* 2016) ranging from LOD0 (least detailed) to LOD4 (most detailed). *LOD0* represents building footprints and Digital Terrain Model (DTM) over which an aerial imagery or a map can be draped; this LOD is a transition from 2D to 3D (Kolbe 2009). *LOD1* is a block model with no

FIGURE 9.2 Example of different LODs according to CityGML classification; the screenshots are taken from the experiment designed in this study: (a) LOD0; (b) LOD1; (c) LOD2; (d) LOD3; (e) LOD3+vegetation; (f) LOD4

FIGURE 9.2 (Cont.)

roof or texture (i.e. extruded building footprints). *LOD2* denotes the 3D block of a building including simple roof structures. *LOD3* includes details such as windows, balconies, doors, and detailed walls and roofs are added to the building façade (Kolbe 2009). At this level, detailed vegetation and modes of transportation such as cars and pedestrians can be added to the model. In *LOD4* interior objects such as rooms and furniture are added to the model (Kolbe *et al.* 2005). The indoor components of buildings are out of the scope of this study because the interior design, other than the internal walls, is not a matter for consideration as part of the DA assessment process. Instead we have tested the influence of vegetation on DA assessments and suggested a new LOD named in this study as *LOD3+vegetation*. This new LOD is similar to LOD3 with the difference that vegetation features are added to the scene.

Virtual reality (VR)

After almost a quarter century of development, VR has become a technology which has been widely adopted for a number of uses; indeed 2016 was acknowledged as the year of VR (Samsung Electronics America 2016; Swanson 2016). As far back as 2001, Manovich (2001) wrote that not only was VR the language of new media it would become 'the accepted way to visualize all information'.

Contemporary VR provides us with highly realistic and dynamic experiences. It provides a high-level representation of spatial objects with high levels of interaction tending towards photorealism (Abdul-Rahman and Pilouk 2007). VR operates fundamentally through two key mechanisms, immersion and presence. Immersion is generally understood as an objective description of the technology that delivers an experience via the senses. The technology is considered more immersive if the experience is more realistic (Slater and Wilbur 1997). Realism is afforded through visual, audio and haptic information. Contemporary VR headsets have up to 100 degrees of horizontal field of view and 110 degrees of vertical field of view; this compares well to the portion of our field of view where both eyes contribute to binocular vision. Some contemporary headsets have built-in stereo headphones and wireless hand controllers, where both make a significant contribution to the VR experience. Presence is characterised as the feeling of being there; as such it can be considered the psychological counterpoint to immersion. There are many theoretical models describing presence, although some use confusing descriptions and attempt to measure in terms of immersion – a classic example was undertaken by Slater and Wilbur (1997) – but there is common agreement that the ability to act in an environment was a key parameter. Riva (2009) outlined the key arguments relating action to presence from the perspective of the cognitive sciences, and concluded that VR research should focus more on providing opportunities for action.

In terms of the built environment many researchers noted that VR environments are able to improve communication and decision-making (Al-Kodmany ; Kjems ; Westerdahl *et al.* 2006). VR facilitates the understanding of a proposed design by clients, both spatially and in use, prior to construction (Morgan and Zampi 1995). Extending this notion, Rosser *et al.* (2007, p.182) found that continued exposure in VR environments results in 'positive benefits … [that include improved] … spatial visualization and mental rotation'. In other words, VR environments don't so much just improve our communication as they seem to improve our *ability* to communicate and our *ability* to understand.

Spatial reasoning, in virtual reality environments, is supported by the facility to compare one's body with objects in space and the ability to compare relationships between different objects (Matlin 2005). Plank *et al.* (2010) described these comparisons as employing egocentric and allocentric reference frames, and stated that both systems interact in the formation of spatial knowledge. Kjems (2005) noted that VR environments inspire dynamic decision-making because of the ease of communication and discussion of alternatives in the shared setting. Roupe and Gustafsson (2013) supported Kjems (2005) in that a key reason for using VR technologies in urban planning is 'access to a shared virtual space, which can facilitate communication and collaboration in order to make better decisions' (2013, p.437), but noted that virtual environments are not without bias, as VR is a representation of space, just as 3D models and 2D perspectives have been in the past. It is unsurprising that agendas inscribed by VR designers may seek to influence the decision-making process. That Roupe and Gustafsson (2013) felt the need to point out the possibility of bias speaks to the compelling nature of VR to convince participants that they are in control of the experience. The LOD experiment presented in this chapter directly addresses detail as a potential source of bias. For example, if a contentious modern development and its heritage zone context were presented as a simple block model, then the contrast between the materiality of the old and the new would be diminished, thereby potentially skewing the decision-making process.

While shifting notions of presence and bias provide theoretical challenges to VR, there are also pragmatic challenges relating to the creation of VR experiences, as well as its use by end users. Many researchers struggled with the creation of real time interactive environments using computer gaming engines. Lowe *et al.* (2011) provided a comprehensive account of these difficulties. Cram *et al.* (2014) undertook a survey which found that the process required to utilise 3D digital models in a computer game environment appears to have lessened the number of geometrical iterations undertaken in the design process. This points to the challenges of working with multiple pieces of digital content creation software (for texture model and audio creation) before processing and combining the resulting assets into a complete environment inside VR world building software. In short, the skillset required to create rich VR experiences is most likely outside those expected of technicians working in real estate and GIS-focused roles. With the increasing popularity of VR, challenges for end users are becoming much easier to overcome. Current high VR headsets, such as the Oculus Rift and the HTC Vive, are tethered by motion restricting wires back to a large and expensive computer. Even with this level of computational power, motion sickness can still be an issue. Companies such as Facebook, Google, Samsung and HTC have embraced VR and have outlined significant milestones in their strategic planning. One emerging technology represents a new category of VR; called 'Standalone VR', these headsets don't require a computer or a mobile phone. They promise to make VR more accessible, more comfortable, more mobile and ultimately more social.

Research methodology

In this study we combined multiple methods including both qualitative and quantitative techniques. As shown in Figure 9.3, in the initial stage the interview questions were designed for conducting focused group sessions to address the first question in this study. DA planners, strategic planners and staff with GIS and 3D modelling expertise were invited to attend these preparatory sessions to (i) identify the preferences and points of view of participants in terms of the evolution of DA assessment processes, and (ii) to test the possibility of incorporating 3D/VR technologies into the DA workflow.

The information gathered from the interviews was used to design an online questionnaire which itself linked varied DA assessment criteria to different LODs. A virtual urban environment was built with different levels of detail using 3D modelling techniques embedded into a virtual reality platform. In this study the experiment is referred to as the 'LOD experiment'. The experiment contained two phases: (i) navigating through the virtual environment; and (ii) completing a post-experience online questionnaire. In this experiment the participants wore an Oculus Rift VR headset for the first phase of the experiment, i.e. for navigating through the virtual environment. The experiment was then followed by the online questionnaire to evaluate the visualisations with varied levels of detail in regards to their perceived views of the performance of the 3D virtual environment with respect to its ability to support specific DA assessment tasks. Information collected from this approach was used to respond to the second research question. The sequencing of these steps is outlined in Figure 9.3.

Build 3D digital models with varied LODs

The VR immersive experience in the LOD experiment was created using *Unreal Engine 4* (version 4.15), which has been developed by computer game developer Epic Games. Typically,

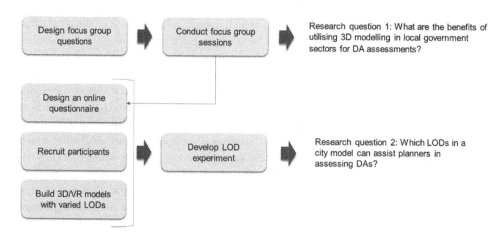

FIGURE 9.3 Design of the study

a designer uses digital content creation tools to author textures, geometry, animations and sound, for example, then exports those assets for subsequent import into a game engine editor. However, Epic Games also supports developers through a market-place where content creators are able to sell digital assets they have created. Often this speeds up the prototyping of a game idea for a developer. In the case of this study the research team purchased an environment that represented a generic virtual city and modified it for use in the experiment. The environment purchased is called Urban City, by PolyPixel; note that a preview can be viewed online at https://youtube/V5VMJmtMyjQ.

Then multiple copies were made and stripped away levels of detail stripped away in each to provide the distinct LODs required for the experiment. Four different LODs were designed for this experiment: LOD1, LOD2, LOD3 and LOD3+vegetation (see Figures 9.2.b–e, respectively). Finally, functionality for users to navigate each LOD environment and navigate between LOD environments was implemented in the VR platform. This functionality was coded via the visual scripting facility inside Unreal Engine 4 called 'Blueprint'.

Participants

Seven professional DA planners, strategic planners and urban designers with GIS and 3D modelling skills attended a focus group session related to the evolution of available methods applied to support DA assessment. The interviewees' selection criteria were based on their professional experience in undertaking DAs or DA-related tasks at three municipalities across Sydney: Blacktown City Council, Randwick City Council and Waverley Council.

In total, 48 participants attended the LOD experiment (n=10 Blacktown Council; n=19 Randwick Council, n=19 Waverley Council). Participants visualised the scenes and ranked their level of understanding corresponding to each LOD. Participants' professional experience in the field of planning ranged from three months to 40 years (Mean=13.1, SD=8.7). Participants were identified as one of the following: planners (n=26), strategic planners (n=10), building surveyors (n=3), landscape architects (n=2), urban designers with GIS 3D skills (n=2), sustainability officer (n=1) or had experience as both DA planners and strategic planners (n=4). Figure 9.4 illustrates the distribution of participants according to their expertise and skills: 52% of the participants were working as a DA planner, 22% had practical

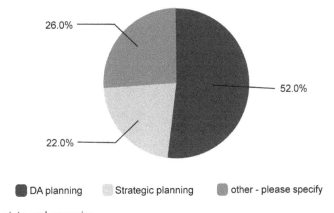

FIGURE 9.4 Participants' expertise

expertise in strategic planning and 26% had other expertise which includes GIS, 3D skills and urban design.

Design the LOD online questionnaire

The LOD questionnaire included two parts. The first part consisted of three general questions which focused on participants' practical experiences in the field of planning. The second part included eight questions about the usability of each LOD for supporting different DA assessment tasks. Specifically, these tasks included the assessment of the following attributes: building height; building setback; building form; shadow; view loss; heritage; urban design; and evidence-based planning. The above-mentioned DA assessment criteria were derived from a review of the information shared in the focus group sessions.

For each DA assessment task, the questionnaire listed the four different LODs available and asked how useful each LOD was to support the completion of that task. Usefulness in the questionnaire was described by a five-point Likert Scale (very useful; useful; moderate; hardly useful; not useful). Figure 9.5 illustrates the online questionnaire for the task: "How useful is each LOD for assessing building setback during DA assessment?"

Since the questionnaire was applied after participants had had the immersive 3D/VR experience, illustrative images of each LOD were attached to each question as a reminder of the environment previously experienced.

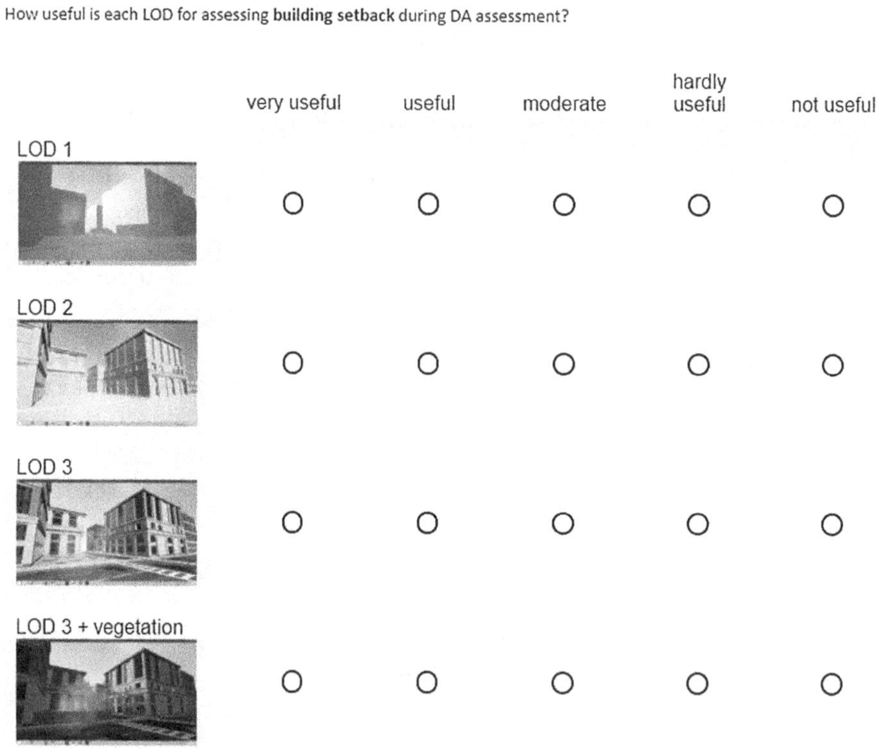

FIGURE 9.5 An example of a question in the online questionnaire; LOD examples are attached to each question

LOD experiment

The LOD experiment was conducted with one participant at a time using Oculus Rift equipment for providing the immersive experience. Three experiments were conducted at the local councils' administrative buildings during three different sessions in May 2017. Metabox prime-X series, i7-6700K CPU @ 4.00 GHz was used to run the program. Participants were briefed on LODs prior to starting the experiment. A test experiment was run to allow participants to become familiar with the VR headset and an Xbox controller. Following the briefing, participants were shown the following instruction on the project information form:

> During the experiment, you will be shown four images of 3D buildings with different LODs, one at a time. These images are taken of a generic city that you will navigate through in this experiment. During this time, please study the image, because afterwards you will be asked to rate the usefulness of each LOD for different building assessment criteria. Next, you will be given an Xbox controller and a Virtual Reality (VR) headset to practice navigating in a Virtual Environment (VE). When you are ready, you can start the experiment by informing the instructor. You will be given up to 5 minutes to navigate through the VE. During this time try changing the LOD using the Xbox controller (up/down on the D pad). You can fly through the VE by pressing the "Y" on the D pad as well as walking through the VE. Once you finish, you can notify the researcher and remove the headset. After the experiment you'll be asked to fill an online questionnaire which would take less than 5 minutes. It will take approximately 10 minutes to complete the whole experiment.

To avoid any occupational health and safety issues, participants were advised to sit down on an ergonomic office chair while they were navigating through the virtual environment. At the end of the experiment, participants were asked to answer the online questionnaire using the provided laptop. Figure 9.6 shows some of the participants during the experiment at the three local councils.

Develop the focus group session questions

A focus group session rather than an individual interview was chosen to provide deeper and richer data through effective means of exploring the participants' thoughts, attitudes and feelings. The interaction between interviewees and the interviewer can generate the synergy and incremental increase in the results (Costigan Lederman 2009).

Accordingly, three sessions were run at the three local councils in April and May 2017, each with a duration of approximately 30 minutes. Interview meetings were arranged with the interviewees at their preferred time. In total, seven experts attended the focus group sessions. Responses to the structured questions were recorded and collected at the same time as the focus group was held. The following questions were addressed in the focus group session:

- *What are the key changes in DA processing from the past to present?*
- *What is the council's vision regarding implementing 3D modelling in DA assessment processes?*
- *What are the benefits of 3D modelling for DA assessments?*

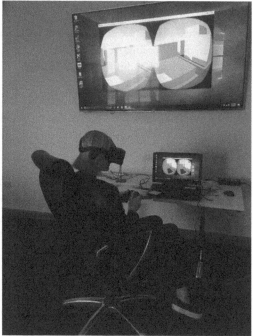

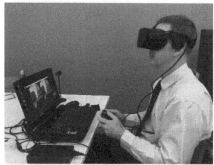

FIGURE 9.6 Participants wearing the Ocolus Rift VR headset and navigating through the VE during the LOD experiment

- *What are the concerns of the council for implementing 3D modelling as a requirement for DA lodgement?*
- *How do you think virtual reality (VR) can be implemented in planning, in general, or in assessing planning proposals, in particular?*

Question 1 aimed to find out the trend of changes on DA assessment from past to the present (i.e. May 2017). Understanding the evolution of the DA process is important when determining the likely benefit 3D modelling can play in the DA assessment process, both current and future. Question 2 illustrated where the current demand was for 3D modelling in the context of local councils across Sydney. In other words, this question was used to gauge feedback on where 3D modelling might be destined to support planners. More specifically, the research endeavoured to find out if 3D modelling supported assessing building height, form and material of the proposed building, visual privacy, views and impact on the foreshore. *Does it help residents to visualise the approved DA?* Question 3 sought to find out the advantages of submitting 3D models of buildings prior to DA approval. The responses to this question could reveal some of the 3D modelling opportunities for local governments in improving the efficiency in the DA assessment process. Question 4 attempted to find out potential issues and concerns regarding the 3D modelling lodgement for DA. It aimed to reduce these concerns and provide some suggestions for future work. While one of the interviewed councils had already added the requirement for lodging 3D digital models for particular development applications and specific sites, the other two did not have this requirement for DA lodgement. At these two councils an applicant can lodge an application online and a planner needs to go through the checklist of the requirements and basic information regarding DAs. For one of these councils an applicant could search the website to access the rules and permissibility using the property address.

Results

Focus group session

Responses to the interview questions by seven senior planning professionals are presented in this section. With respect to Q1 the main changes mentioned by the participants were related to the introduction of digital maps and computer systems to support DA assessment tasks. In general these changes improved the performance of the procedures via more accurate maps, more efficient systems to overlay those maps, and also to produce and record amendments and to back up materials. Searching information to support DA assessment and delivering information to people potentially interested (e.g. DA tracking) were also improved through online services and mobile phone apps. The use of 3D digital models in DAs is not a current practice yet.

Table 9.1 provides a comprehensive summary of responses to the first interview question. Prior to having access to basic digital maps, local councils used to have hand-drawn maps which were not very accurate. In 1985, Local Environmental Plan (LEP) maps were the only references to planning controls available in Sydney. These maps were hand-drawn and hand-coloured. Councils had few copies of these maps, which were usually hung on the walls in council offices. Any amendments to the maps were very difficult to perform. Basically, a sticky Post-it note used to be stuck on the amendments to update the map. Accordingly, the maps were not very reliable and accurate. In this context, maintaining fundamental DA-relevant information was limited, as it was not possible to back up any maps rather than keeping hard copies. With the emergence of digital maps as supported by geographical information system (GIS) technology, the possibility to back up the digital maps was provided. In the past it was impossible to overlay physical maps, but the emergence of digital maps solved this short-coming. Searching for information was difficult and time-consuming using paper maps. Today there is a myriad of different software to make it easier to search for property addresses and their required information. DA tracking and mobile apps are just some examples that help residents to check and track approved DAs (for example see http://planning.randwick.nsw. gov.au/pages/xc.track.advanced/searchapplication.aspx).

The research undertaken found the use of 3D models – either a physical model or a digital 3D model – were not requirements for lodging and assessing DAs. Some councils have used physical 3D models in the past, but since they were bulky, time-consuming to build and costly to maintain, they were not commonly used. With modern technology such as GIS, CAD and BIM, it is now possible to readily create and use digital 3D models to support the DA process. However, this research found that there is limited utility of 3D models in planning practice in the context of Sydney, except for the occasional use for assessing building height controls.

Previously, DA assessment mainly relied on information provided by the applicant that mostly used photo-montage methods, so shadow analysis, for instance, was checked using hand drawings. Nowadays, there are automated methods for conducting shadow assessment for DAs. However, applicants typically submit to councils 2D maps and drawings in pdf format, which do not enable planners to interact with the design. Hence, shadows are still drawn by hand and if they do not look correct applicants have to change them manually.

TABLE 9.1 Changes of DA processing from past to present

Changes	Past	Present
Accuracy of maps	Hand-drawn and hand-coloured maps with low accuracy	Digital maps with improved accuracy
Plan amendments	LEP map was posted on the wall and plan amendments were updated on the map by sticking new paper over the old map.	Amendments are made digitally on maps using appropriate computer mapping systems such as GIS
Back-up	Not available	Ability to back up digital maps
Map overlays	Used separate maps for different attributes (e.g. heritage, stormwater map), difficult to access from different departments, and also difficult to be visually integrated	Digital map layers are easily overlaid, analysed and visualised
Using 3D models	Sometimes physical models were built, but they were bulky and required too much space, particularly for large scale developments	Still in its infancy, the use of 3D digital models in some councils is supporting building height controls assessment
Searching for information	Searching for detailed information on paper LEP maps was a very time-consuming task	Online systems are now available for easy and quick search by property address and using pathways
Open access DA tracking	Not applicable	Ability to track approved DA
Mobile applications	Not applicable	Local residents can get notifications of DAs in their locality through council applications
DA enquiries	Applicants needed to contact duty planners	Online DA enquiries make it easier for applicants prior to lodging DAs
Electronic DA submissions	Hard copies (usually six A0 size plans)	Electronic DA submission (usually in pdf format)

The respondents indicated that changes in the DA process mostly occurred during the assessment stage, which typically ranges from 30 to 70 days. This is the most important stage of the DA process which results in the DA outcome, i.e. either approval or rejection. Although DA assessments have improved dramatically from the past, planners still need to spend a considerable amount of time on information checklists. It is anticipated that it is at this stage when 3D modelling could make a significant contribution. In addition, the direct outcome of the interviews with planners was that 3D models are not commonly utilised by planners, and particularly for DA assessments. There are very few cases in which the above-mentioned councils used digital 3D models, and when this occurred it was mainly for evaluating height controls and court cases. There may be two possible reasons for this situation: first, 3D digital modelling software by itself cannot assist planners to assess DAs, a fact that is contrary to the assumption of the present study; and secondly, there exists a lack of training in certain software which limits the applicability of 3D digital models at local councils.

The second question discussed in the focus group session sought the councils' vision in implementing 3D digital modelling in DA assessment, and to find out the current demand for

3D modelling by local governments. The vision of one of the councils was to have a 'fast holistic and accurate assessment of DAs using 3D technology to guide our[the] planning decisions and allow us[planners] to test and trial different built forms before making recommendations or conditions to development…'.

The respondents indicated that the current demand for 3D digital modelling in the interviewed councils was for:

• Additional tools for better understanding spatial dimensions of buildings, arrangements of buildings, scale, measuring room sizes and to check solar access in DA assessments;
• Automated methods in assessing the applications to validate the model; for instance, checking whether the proposed model complied with the current controls, check the contours and so forth;
• Additional tools for facilitating court cases; for instance, the height plane could be overlaid on the 3D model of the applicant to show where the height exceeds the controls;
• Additional tools for facilitating and communicating ideas with applicant, particularly for pre-DA lodgements, e.g. show them the 3D models with different height and the impact of them;
• Mapping the entire municipality in 3D and input new buildings into the software to improve the assessment process;
• Open access to the 3D models as part of the DA tracking process, from submission to approval and then beyond; for reuse for other purposes.

According to the participants, 3D modelling was suggested to be implemented across council not only for DA assessments, but also for a wider range of purposes including: resolving disputes among neighbours; at court for better understanding of the impact of a design; for the formation of development controls; and interior building design, which can measure room sizes and modify windows or door locations.

The third interview question endeavoured to uncover the implications of 3D modelling for local governments, particularly for DA assessment. The following list of responses resulted from the focus group session and included assessment tasks related to the DA approval process that could be assisted by 3D digital modelling and visualisation:

• Shadow analysis (planners wish to use 3D modelling to calculate overshadowing impacts);
• View shed, vista assessment and view loss;
• Height analysis;
• Form analysis e.g. floor space ratio (FSR), setbacks and balcony analysis;
• Understanding the orientation of the buildings;
• Urban design and built form analysis;
• Streetscape and context analysis, e.g. accumulative impact; how the development impact the streetscape and shadowing;
• Heritage perspectives, e.g. how development affects heritage area;
• Colour, material and finishes, e.g. whether a proposal is compatible with heritage or foreshore Sydney protection areas;
• Improve relationships with public domain;
• Community awareness, i.e. require less time from planner to explain design proposals for new developments to the community and solve related disputes;

- Better communication between council staffs, e.g. DA planners, transport engineers, etc.;
- Zonings and planning controls, e.g. DCP;
- Visualise and test different building envelopes;
- Fast and accurate information, e.g. '3D cannot hide anything', quoting from one of the council interviewees;
- Court cases, e.g. 3D can be used for discussion at court as evidence-based planning;
- Test and view different building envelopes;
- Topographic analysis; and
- Strategic planning; 3D can guide in strategic planning decisions and improve relationships with public domain.

It was found that wind assessments of new developments are currently the responsibility of applicants using wind tunnel exercises. However, if 3D models provide the opportunity for planners to assess the wind, there is the possibility of councils validating submitted wind assessment. Visual impact analysis of façades was another assessment that councils wish to pursue, but due to the limited facilities and the complexity of the analysis, they cannot easily undertake.

The fourth question investigated the concerns of local government in implementing 3D modelling as a requirement for DA lodgement. There are some questions that council members are struggling with, such as whether they need to charge the applicants for lodging the models. Another concern of the councils was the required training for current staff in order to work with 3D plans. They also preferred to have an existing conditions model available and replace the 3D model of the new development with the uploaded model of the applicant to enable a comparative analysis of current and proposed future development. Some suggestions were provided to address the concerns raised at the focus group session/s regarding implementing 3D modelling, as summarised in Table 9.2.

TABLE 9.2 Local government concerns and the authors' suggestions for implementing digital 3D modelling for DA assessment

Concerns	Suggested mitigations
Councils need to get effective 3D models for their demands, which could be easily used and understood by their planners	*Accuracy of models/plans* – procedures and specification can be set for applicants to submit the model
Councils and their selected 3D models must have the ability to adapt to technology changes and not have a short life of only a couple of years	*An integrated online platform* – integration of 3D modelling and DA through an online platform which could automate and speed up the basic information check of DAs
Cost and maintenance of a 3D model may be a limitation for its implementation	*Financial applications* (i.e. cost for councils) – council should consider costing in the requirement for 3D model submission and evaluation into the DA process
Resources, including software, staff and suitable work stations, are required	*Resource* – providing software, staff members, computers. Automated check of the models can be provided online. Customers can ask any questions before submitting the model rather than changing it later

Finally, the fifth question sought to evaluate the potential for applying VR in assessing planning proposals and DAs. The interviewees stated the perceived benefits of VR for understanding proposals by navigating through the realistic virtual environment and around the proposed development. VR has the potential to make it possible for planners to manipulate the design, for instance the height of a building, which would be most beneficial in the assessment process. Facilitating the sense of scale is another advantage of using VR in assessing DAs which can lead to a better understanding of the built form.

Besides the advantages that VR technology could bring for council tasks, the interviewees preferred to have the on-screen 3D models rather than using the VR headset. The cost of VR was the largest deterrent factor for not supporting the use of this technology for DA assessment. The last concern for some of the interviewees were that wearing the VR headset made them feel dizzy and suffer motion sickness.

In response to the motion sickness issue, there are various methods that can help reduce this problem, such as matching the VR to the real world as much as possible (Allen *et al.* 2016). Since developers are focusing on reducing the motion sickness to zero, 'The VR industry is moving at a pace science can't match, forging ahead with its own grand experiment as millions of users test its products' (Mason 2016). Regarding the cost of VR, it can be argued that mistakes in assessment could be more costly than implementing VR. Given the benefits of VR for urban planners, this technology is suggested for assisting DA assessment processes.

LOD experiment

In this section the results of the LOD experiment are reported, based on the responses from the online questionnaire associated with the experiment with the 3D models in virtual reality. Respondents assigned a rating (within a five-point scale from 4/very useful to 0/not useful) for each LOD (from LOD0/Most simple to LOD3+vegetation/Most realistic) for eight different tasks (outlined previously) involved in DA assessments. Figure 9.7 summarises the responses from all 48 participants indicating their proportional preferences. Each graph represents the results for a specific DA assessment task. The percentage of frequencies of responses to each task is calculated. In the graph the y axis represents the percentage of responses to each LODs, the x axis represents the 'usefulness' of each LOD for the specific DA task and the z axis indicates the proportion of participants' preferences.

These results highlight the 'very useful' category in Figure 9.7, which indicated that the most realistic 3D model in virtual reality (LOD3+vegetation) was selected for all DA assessment tasks. However, it also shows that for some tasks LODs with lower levels of details can perform similarly to LOD3+vegetation – e.g. LOD3 for building setback and building form, or LODs 2 and 3 for shadow.

In order to ease the comparability among LODs for each DA assessment task, a weighted average of 'usefulness' rank has been assigned to each LOD. This weighted average is based on frequency of responses with the usefulness as weights, ranging from 4 (very useful) to 0 (not useful). Moreover, to provide an overall rank for each LOD, the combined averages from all DA assessment tasks were calculated. The results, presented against the five-point scale for usefulness, are shown in Figure 9.8.

Overall LOD1 was the lowest-performing 3D urban model to support DA assessment. Participants assessed the usefulness of LOD1 as 'Moderate' for almost all DA assessment tasks, with the exception of heritage, for which it was classified as 'Hardly useful'. For instance,

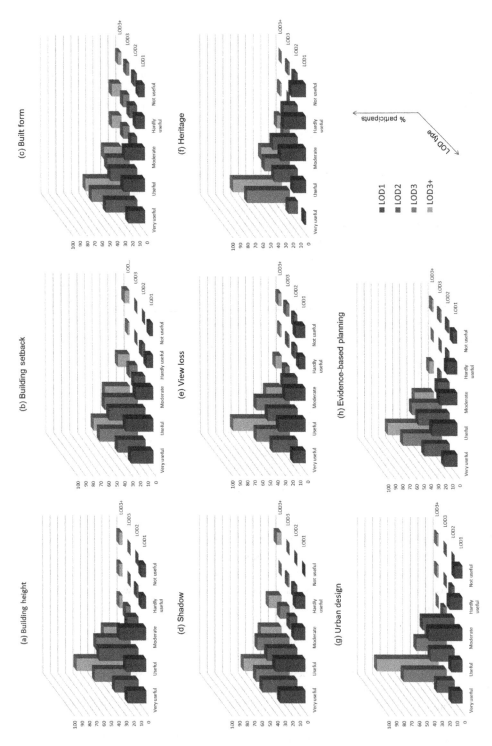

FIGURE 9.7 Results from the LOD experiment: participants' evaluation of usefulness of different LODs for DA assessment tasks

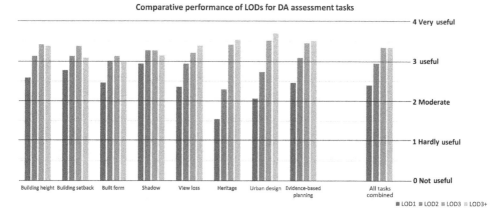

FIGURE 9.8 Comparative usefulness of LODs for DA assessment tasks and combined averages from all DA assessment tasks

LOD1 can be mostly used for building setback assessments, built form analysis, shadow analysis, as a tool at court cases (e.g. for evidence-based planning), and finally to evaluate view loss. LOD1 is not practical for heritage analysis, as the texture and materials of buildings are needed. This statement is also relevant for urban design analysis.

LOD2 was considered 'Useful' for the majority of DA assessment tasks and 'Moderate' for few tasks (heritage, urban design, particularly). This level of detail can be used for evidence-based planning, building height and setback assessments, as well as evaluating view loss. However, because there is not much detail of the façade available, it is not very useful for urban design and heritage conservation studies.

LOD3 and LOD3+vegetation were the highest-performing models. Despite having the same overall results, they differ slightly for some tasks. For DA tasks such as the assessment of building height, setback, form and shadow, LOD3 performed better than LOD3+vegetation; for other tasks, such as the assessment of view loss, heritage and urban design, LOD3+vegetation performed better than LOD3. These results suggest that vegetation does not play a significant role in assessing the above-mentioned tasks. However, the presence of vegetation can help planners in better performing other tasks, such as shadow analysis, building height and evidence-based planning, since vegetation can work as a reference for scale and view loss, for example.

Conclusions

This chapter examined two lines of research that have not previously received much attention. First, the implementation of 3D digital modelling and VR for DA assessments for municipalities. Secondly, the assessment of LODs and their perceived utility in supporting DA-related tasks. In this process, two research questions were identified. The first question sought to find the benefits of 3D modelling and VR for DA assessments and whether 3D modelling can ease and enhance DA assessments undertaken by municipalities. To evaluate this question three separate focus group sessions were conducted across three local government councils in the Greater Sydney region. Table 9.3 summarises the opportunities and

TABLE 9.3 Potential opportunities and barriers of implementing 3D VR models in DA assessments

Opportunities/benefits	Barriers
Accurate building models	Cost and maintenance of a 3D model and VR equipment
Accurate façade details	Resources, including software, staff, and suitable work stations
Overlay building data	
Assigning building controls	
Visualise approved DAs	
Accurate analysis such as shadow, line of sight, visual impact analysis, etc.	
Ease of DA submission	
Interactive format	

barriers of using 3D in DA assessment resulting from these sessions and addresses the first research question.

The second research question referred to which LODs in a 3D digital model can assist planners to assess DA in a faster and more efficient way. The LOD experiment was designed to respond to this question and measure the efficiency of different LODs for DA assessments. This research investigated the usefulness of each LOD for DA and also planning proposals assessments. Results from the experiment show that participants were keen on using the 3D digital modelling not only for DA assessments but also for community engagements, streetscape and urban design related projects, as well as interior design exploration (i.e. LOD 4, which was outside the scope of this research). Additionally, evidence suggests that LOD 3+vegetation is the most useful LOD level of assessment. This supports the necessity of including vegetation in proposed changes of a building, as it is currently being assessed separately as part of the landscaping and open space assessment. In fact, vegetation can help to realise the scale of the model and can be used to complement several additional assessments such as view shed and visibility.

In conclusion, this research recommends the implementation of 3D digital modelling to assist DA tasks and other council-related applications. As per the responses from participants, training and upskilling of council workers in 3D technologies is essential to fully embrace 3D digital modelling in the current urban planner's toolkit. Finally, despite the participants' perceived utility of 3D digital modelling to support a number of DA-related tasks, they were less positive regarding the implementation of the VR technology, mainly because of its cost and VR-related motion sickness. This was somewhat surprising given the recent revival of VR as a platform for representing 3D models and worlds. However, as 3D digital models are not fully utilised in planning, revisiting VR and also augmented reality technologies might be worth perusing once 3D digital models become a ubiquitous technology supporting city planning.

Acknowledgements

We would like to thank Blacktown City Council, Randwick City Council and Waverley Council for participating in the study.

References

Abdul-Rahman, A. and Pilouk, M. (2007). *Spatial Data Modelling for 3D GIS*. Berlin, Heidelberg: Springer-Verlag Berlin Heidelberg. Retrieved from: dx.doi.org/10.1007/978-3-540-74167-1.

Al-Kodmany, K. (2002).Visualization tools and methods in community planning. From freehand sketches to virtual reality. *CPL Bibliography*, 17(2), 189–211; 10.1177/088541202762475946.

Allen, Brian, Hanley, Taylor, Rokers, Bas and Green, C. Shawn (2016). Visual 3D motion acuity predicts discomfort in 3D stereoscopic environments. *Entertainment Computing*, 13, 1–9; 10.1016/j.entcom.2016.01.001.

Biljecki, F., Ledoux, H. and Stoter, J. (2016). An improved LOD specification for 3D building models. *Computers, Environment and Urban Systems*, 59, 25–37. Retrieved from: www.sciencedirect.com/science/article/pii/S0198971516300436.

Biljecki, F., Ledoux, H., Stoter, J. and Zhao, J. (2014). Formalisation of the level of detail in 3D city modelling. *Computers, Environment and Urban Systems*, 48, 1–15.

Birkin, M., Clarke, G., Clarke, M.P. and Wilson, A.G. (1996). *Intelligent GIS. Location Decisions and Strategic Planning*. London: Wiley.

Boguslawski, P. and Gold, C. (2009). Construction operators for modelling 3D objects and dual navigation structures. In William Cartwright, Georg Gartner, Liqiu Meng, Michael P. Peterson, Jiyeong Lee and Sisi Zlatanova (eds.), *3D Geo-Information Sciences*, 47–59. Berlin, Heidelberg: Springer Berlin Heidelberg.

Costigan Lederman, L. (2009). Assessing educational effectiveness. The focus group interview as a technique for data collection. *Communication Education*, 39(2); 10.1080/03634529009378794). Retrieved from: http://nca.tandfonline.com/loi/rced20.

Cram, A., Lowe, R. and Lumkin, K. (2014). Assessing spatial design in virtual environments. In Robert Tennyson, Shannon Kennedy-Clark, Kristina Everett, Penny Wheeler (eds.), *Cases on the Assessment of Scenario and Game-Based Virtual Worlds in Higher Education* (Advances in Game-Based Learning), 74–123. IGI Global.

Dodge, M., Doyle, S., Smith, A. and Fleetwood, S. (eds.) (1998). Towards the virtual city. *VR and Internet GIS for Urban Planning. Virtual Reality and Geographical Information Systems Workshop*. Birkbeck College, London. Retrieved from: www.casa.ucl.ac.uk/martin/tmp/html_version/vrcity.html.

ESRI (2014). CityEngine Example: Redlands Redevelopment, 1–18. Retrieved from: www.esri.com/~/media/Files/Pdfs/library/brochures/pdfs/cityengine-example-redlands.pdf.

Isikdag, U., Horhammer, M., Zlatanova, S., Kathmann, R. and van Oosterom, P.J.M. (eds.) (2015). Utilizing 3D building and 3D cadastre geometries for better valuation of existing real estate. *Proceedings of FIG Working Week 2015 'From the wisdom of the ages to the challenges of modern world'*, 17–21 May, Sofia, Bulgaria: International Federation of Surveyors (FIG). Retrieved from: https://repository.tudelft.nl/islandora/object/uuid:4253db43-a734-4641-ab6b-cd13742ec279?collection=research.

Isikdag, U. (2015). The Future of Building Information Modelling. BIM 2.0. In Umit Isikdag (ed.), *Enhanced Building Information Models*, 13–24. Cham: Springer International Publishing.

Kjems, E. (2005).VR applications in an architectural competition. Case: House of Music in Aalborg. *Realitat Virtual a l'Arquitectura i la Construcció,* Taller 2. *Khora ll*, 47–58.

Klosterman, R.E. and Pettit, C.J. (2005). An update on planning support systems. *Environment and Planning B: Planning and Design*, 32(4), 477–484; 10.1068/b3204ed.

Kolbe, Thomas H. (2009). Representing and exchanging 3D City Models with CityGML. In William Cartwright, Georg Gartner, Liqiu Meng, Michael P. Peterson, Jiyeong Lee and Sisi Zlatanova (eds.), *3D Geo-Information Sciences*, 15–31. Berlin, Heidelberg: Springer Berlin Heidelberg.

Kolbe, T.H., Gröger, G. and Plümer, L. (2005). CityGML: Interoperable Access to 3D City Models. In Peter van Oosterom, S. Zlatanova and E.M. Fendel (eds.): *Geo-Information For Disaster Management*, 883–899. Berlin, Heidelberg: Springer Berlin Heidelberg.

Lin, H., Zhu, J., Xu, B., Lin, W. and Hu, Y. (2009). A virtual geographic environment for a simulation of air pollution dispersion in the Pearl River Delta (PRD) Region. In William Cartwright, Georg Gartner, Liqiu Meng, Michael P. Peterson, Jiyeong Lee and Sisi Zlatanova (eds.): *3D Geo-Information Sciences*, 3–13. Berlin, Heidelberg: Springer Berlin Heidelberg.

Lowe, R., Cromarty, J. and Goodwin, R. (2011). Real time modelling. A solution for accurate, updatable and real-time 3D modelling of as-built architecture. In C.M. Herr, N. Gu, S. Roudavski and M.A. Schnabel (eds.): *Circuit Bending, Breaking and Mending. 16th International Conference on Computer-Aided Architectural Design Research in Asia*, 219–228. Hong Kong: Association for Computer-Aided Architectural Design Research in Asia (CAADRIA).

Luo, Y., He, J., Liu, H. (2015). Application of a 3D modeling software – Cityengine in urban planning. In Ai Sheng (ed.), *Energy, Environment and Green Building Materials*, 455–458. CRC Press.

Manovich, L. (2001). *The Language of New Media*. Cambridge, Mass., London: MIT Press (Leonardo).

Mason, B. (2016). Virtual reality has a motion sickness problem. People prone to nausea may opt out of immersive experiences. *ScienceNews:Magazine of Society and the Public*. Retrieved from: www.sciencenews.org/article/virtual-reality-has-motion-sickness-problem.

Matlin, M.W. (2005). *Cognition*, 6th edition. J. Wiley and Sons.

Millar, S. (2015). Virtual reality OFIs a real estate game changer. realestate.com.au. Retrieved from: www.realestate.com.au/news/virtual-reality-ofis-a-real-estate-game-changer/.

Morgan, C.L. and Zampi, G. (1995). *Virtual Architecture*. New York: McGraw-Hill.

NSW Department of Planning and Environment (2017). Assessment System. Retrieved from: www.planningportal.nsw.gov.au/understanding-planning/assessment-systems.

Plank, M., Müller, H.J., Onton, J., Makeig, S. and Gramann, K. (2010). Human EEG correlates of spatial navigation within egocentric and allocentric reference frames. In Christoph Hölscher (ed.), *Spatial Cognition VII. International conference, Spatial Cognition 2010*, 6222, 191–206.

Randwick City Council (2017). Development Application (DA) process. Retrieved from: ww.randwick.nsw.gov.au/planning-and-building/development-application-da-process.

Riva, G. (2009). Is presence a technology issue? Some insights from cognitive sciences. *Virtual Reality*, 13(3), 159–169; 10.1007/s10055-009-0121-6.

Rosser, J.C., Lynch, P.J., Cuddihy, L., Gentile, D.A., Klonsky, J. and Merrell, R. (2007). The impact of video games on training surgeons in the 21st century. *Archives of Surgery* (Chicago, Ill.: 1960) 142(2), 181–6; discusssion 186; 10.1001/archsurg.142.2.181.

Roumpani, F. (2013). Developing classical and contemporary models in ESRI's City Engine. Centre for Advanced Spatial Analysis (UCL working papers series, paper 191).

Roupe, M. and Gustafsson, M. (2013). Judgment and decision-making aspects on the use of virtual reality in volume studies. In Rudi Stouffs (ed.), *Open Systems. Proceedings of the 18th International Conference on Computer-Aided Architectural Design Research in Asia (CAADRIA 2013)*, 437–446. Singapore: National University of Singapore.

Samsung Electronics America (2016). 2016: The Year Of Virtual Reality. Retrieved from: www.sponsorship.com/iegsr/2016/03/28/2016--The-Year-Of-Virtual-Reality.aspx.

Slater, M. and Wilbur, S. (1997). A framework for immersive virtual environments (FIVE). Speculations on the role of presence in virtual environments. *Presence: Teleoperators and Virtual Environments*, 6(6), 603–616.

Swanson, J. (2016). 2016: The Year of Virtual Reality (knowledgeWorks). Retrieved from: http://knowledgeworks.org/worldoflearning/2016/05/virtual-reality-2016/.

Underwood, J. and Isikdag, U. (eds.) (2010). *Synopsis of the Handbook of Research on Building Information Modelling*. CIB World Congress, Salford, 10–13 May. Retrieved from: www.irbnet.de/daten/iconda/CIB18807.pdf.

Valero, E., Adan, A. and Bosched, F. (2015). Semantic 3D reconstruction of furnished interiors using laser scanning and RFID Technology. *Journal of Computing in Civil Engineering*, 30(4); https://ascelibrary.org/doi/pdf/10.1061/%28ASCE%29CP.1943-5487.0000525.

Westerdahl, B., Suneson, K., Wernemyr, C., Roupé, M., Johansson, M. and Martin A. (2006). Users' evaluation of a virtual reality architectural model compared with the experience of the completed building. *Automation in Construction*, 15(2), 150–165; 10.1016/j.autcon.2005.02.010.

Xu, X., Ding, L., Luo, H. and Ma, L. (2014). From building information modeling to city information modeling. *BIM Cloud-Based Technology in the AEC Sector: Present Status and Future Trends*, 19, 292–307. Retrieved from: www.itcon.org/2014/17.

10

CONCLUSION

Richard Reed and Chris Pettit

Introduction

The objective of this book was to bridge the gap between the property and GIS disciplines, which has been achieved through addressing research problems and providing viable solutions to research questions. This research gap has been previously acknowledged, with the analysis in each chapter undertaken from either a real estate or GIS perspective to highlight the various contributions of each discipline (Dolan *et al.* 2017). Both property and GIS coexist in research, government and industry sectors, even though the synergy between them had not been fully investigated to date. An interdisciplinary approach for understanding the city which draws upon the discipline of property and GIScience does not exist currently, although it is a rapidly emerging area. The increasing availability of location-based information in the 21st century, including Census data, cadastre data and other spatial-related data, at a low or zero cost, coupled with access to user-friendly spatial software programs, have greatly reduced the barriers for stakeholders to use GIS in analysing and mapping our rapidly urbanising world. Previously this was largely restricted to GIS experts and large government and industry organisations. The ubiquitous nature of computing and access to a plethora of open data in many parts of the world highlight the high relevance and ideal timing of this book.

The property discipline has for many years had strong established links with spatial data, and most forms of property include reference to some component of land based on a spatial characteristic (Reed *et al.* 2014). Exceptions to this statement include intellectual property, goodwill and other intangible forms of property. However, these assets only form a minor proportion of the property discipline and are often associated with the law discipline more than the property or real estate discipline. As the cadastre is a comprehensive register of the real estate or properties' spatial boundaries, there is an inextricable link to spatial database, analytical and visalisation functions inherent within a GIS. Low-density housing and even medium- and high-rise condominiums have a spatial location associated with a street address, as do air rights and underground tunnels. However, many property professionals, whether in their formal education or informal industry experience, have no or very limited knowledge about aspects of spatial data and the underpinning GIScience discipline despite the large influence on each property. *Why does this occur?* A large barrier has been the resistance to change in the property and real estate profession from a global perspective (Jylha *et al.* 2014). The operation of the property market, commencing with the transfer process and scaling up to the valuation process, has changed little in many decades and digital modernisation has been limited. Whilst there has been progress in real estate agents advertising via the internet and also internet-based title (or deed) transfers, for example, most of the property-related tasks involve human interaction and judgement with minimal workflow automation and little use of spatial analytics.

At the same time, there has been enormous progress within the field of GIScience with growth in available GIS software, both proprietary and open access. Geospatially enabled technologies are also becoming increasingly easy to access, via both desktop and cloud-based software platforms. Recently there has been a reawakening in virtual reality and augmented reality as supported through the use of gaming engines. These hold promise for the new and engaging immersive tools to engage those in the property and real estate sectors. With new disruptive technologies making their way on to the global stage, like many sectors the real estate industry faces challenges in being upskilled with current data analytic skills to undertake both operational and strategic tasks.

The challenge for the spatial industry is the ability to engage deeply into domain-specific areas such as property and real estate. It is promising to observe location intelligence being embedded into a number of property search platforms and unmanned aerial vehicles (UAVs) which are increasingly being used to provide virtual property tours from a height. However, there exists much untapped spatial analytical functionality that the real estate industry and property sector could benefit further from. In this book a number of more advanced spatial analysis and visualisation methods have been applied to address important research questions. Unfortunately it has not possible to encompass the full breadth of what GIS can offer for every scenario in the property and real estate industry due to the resource limitations of one book. For example, there are substantial opportunities to tap into building information models (BIM) and smartphone applications which utilise spatial intelligence and geoportals such as the Australian Urban Research Infrastructure Network (AURIN) (Pettit *et al.* 2017).

From the perspective of the GIS profession, the chapters in this book illustrated just some of the potential applications of spatial technologies and methods within the context of the property and real estate industry. Such applications which may be of interest to property and real estate professionals would potentially also be of interest to those who are tasked in city shaping; for example, stakeholders including urban- planners and policy-makers. As we are living in an increasingly urbanised world, the use of spatial information and spatial technologies assume an increasingly important role in ensuring humans live in sustainable, productive and resilient cities (Geertman *et al.* 2017). The next section summarises the outcomes from the previous chapters and discusses each contribution to the broader property and real estate context.

Chapter findings and contribution

A summary of the findings from each chapter is listed here.

Residential intensification and housing demand: A case study of Sydney, Brisbane and Melbourne (Chapter 2)

Australian cities are facing significant population growth, with both Sydney and Melbourne each forecast to accommodate more than eight million residents by 2050, followed by Brisbane and Perth with each forecast in the vicinity of five million residents by the same time. Due to such rapid growth, Australian cities are facing housing intensification and there is a need to inform stakeholders using compact city policies. This research chapter built upon previous work published by Randolph and Tice (2013) and extended this method to examine the demand drivers for higher density living across three Australian cities Sydney, Melbourne and Brisbane. These cities are recipients of the highest levels of housing intensification across Australia.

This research addressed two key questions. First, *What are the characteristics of the demand drivers for high-rise living for Sydney, Melbourne and Brisbane?* Secondly, *How are the different groups accommodated by this market distributed across the three cities?* The 2001 Australian Bureau of Statistics Census data provided the necessary data (48 socio-demographic variables) with which to profile households living in multi-unit housing. Households were analysed in Statistical Area 1 (approximately 200 households) using a Principal Component Analysis (PCA) method with the results displayed as maps using a GIS. The resultant map layers have also been made available through the City Futures' CityData platform (see https://citydata.be.unsw.edu.au/).

The findings from this research showed that in Sydney the key components were (i) Overseas students and 'Generation Y' renters, (ii) The economically engaged, (iii) Multicultural family households and (iv) Lower income retirees. For Melbourne the results were somewhat similar with the key components being (i) Overseas students and 'Generation Y' renters, (ii) The economically engaged, and (iii) Multicultural familty households. However in the case of Melbourne there was no fourth component observed. For Brisbane the key components were (i) The economically engaged, (ii) Lower income retirees and (iii) Low income. Interestingly, across both Sydney and Melbourne the first three components were the same, with the most prominent component being the Overseas students and Gen Y renters. Brisbane components were somewhat comparable but less distinctive.

Through the use of GIS, the spatial distributions of the five sub-markets across the three cities have been visualised through a series of maps. Interestingly, the economically engaged component confirmed a strong spatial clustering existed in and around the respective city CBDs. Other key components were not so centrally clustered and extended to pockets in the middle and outer parts of these three cities. The results from this research are of direct interest to the real estate industry, city-planners and policy-makers, as they highlight distinct higher density living market segments which exist across the three cities.

How disruptive technology is impacting the housing and property markets: An examination of Airbnb (Chapter 3)

In an era of pervasive computing there are a number of disruptive technologies which are having widespread impact on cities. Most notable are property rental matching services such as Airbnb and mobility services such as Uber. In this chapter the authors looked critically at the spatial and regulatory impact of Airbnb on three large cities across three continents, specifically, London, Sydney and Phoenix, for the basis of three comparative case studies. Fine scale property level Airbnb data has been acquired from the company AirDNA and has formed the basis for a number of spatial analysis methods which have been deployed to understand the likely impact of Airbnb on the three cities studied. The spatial analysis techniques used include: dot density mapping, buffer analysis, Moran's *I* and location quotient analysis.

In this chapter the aim was to explore how spatial data analysis tools can help to support cities in responding to the rise of Airbnb in an equitable and efficient way. The research was primarily focused on understanding how the balance of whole-unit to shared-unit Aribnb listings varied by area, and to benchmark this against the metropolitan region as a whole, because different cities have different overall numbers of Airbnb listings. For example, London and Sydney had approximately 22,000 listings each, while Phoenix had approximately 6,000 listings. The analysis undertaken in this research focused on market segmentations of Airbnb properties broken down into the categories (i) traditional holiday lets, and (ii) multi-listings, which were the focus since they provide a good proximity for properties which are likely to be impacting the long-term rental stock within a city. The portions of shared properties were also investigated. With respect to greater metropolitan London, Sydney and Phoenix, these vary considerably in size and structure. London and Sydney have more defined central business areas compared with Phoenix, which is more of a polycentric metropolitan region city. The findings of this research show there are more Airbnb listings closer to central London, then Sydney and then downtown Phoenix, which aligns with the compactness of the respective metropolitan areas.

Overall for Sydney the data indicated that approximately 30% of Airbnb listings can be defined as a traditional holiday let and similarly 30% used as multiple listings. With respect to multiple listings the trend is even more pronounced with London and Phoenix, for both metropolitan regions have over 40%. In the case of London, what is worth noting is that Airbnb could be having an impact on the gentrification of historically deprived inner neighbourhoods such as Hackney.

The research undertaken in this chapter illustrates how GIS methods can be applied for understanding the spatial patterns and distribution of rental accommodation matching platform, Airbnb. Through the use of visualisation tools such as dot density maps and more sophisticated spatial statics such as Moran's I, the hot spots of activity of Airbnb can be mapped and analysed. This research provides important and timely research to support city-planners and policy-makers in developing appropriate regulatory environments to deal with the challenges and opportunities posed through disruptive technologies and platforms such as Airbnb.

The contribution of GIS to understanding retail property (Chapter 4)

This chapter focused on retail land use and demonstrated how GIS can provide invaluable insights into a retailer's location and the distribution of purchasers. The scope of the research was limited to 'retail destinations', being an industry term referring to enclosed shopping centres, regional shopping centres, large format retail centres or bulky goods centres; the term also encompasses other forms of retail centres including 'high street' or main street/strip shopping precincts. A list of retail destinations is included in Appendix B. A fundamental characteristic of retail destinations is to determine the geographical catchment area for prospective purchasers. In other words, it is essential to assess which purchasers would prefer to travel to a particular retail destination rather than to travel to competing retail destinations. This task is further complicated for prospective new retail destinations where there is no trading history or customer base to examine. Therefore the research question for this chapter was: *What are the drivers linked to retail customers visiting a specific retail destination rather than a competing destination?*

This analysis was based on a case study approach and, although the application of GIS has implications for varying types of retail land use, this case study was designed for use with larger retail holdings. The approach was to use GIS to identify potential scenarios for locating a new large format retail store in a regional city. The findings confirmed GIS was able to facilitate identification of multiple potential factors which influenced the success of the retail destination related to the location and demographic characteristics of existing households, the potential catchment area and the location of competing retail destinations. In addition there were other locational factors which contributed to the analysis; these included the trade area, local planning considerations, as well as highest and best use of the subject site. The findings confirmed a detailed GIS analysis was essential in an effective retail land use evaluation.

Modelling value uplift on future transport infrastructure (Chapter 5)

Property land valuation and the potential changes in property value are important considerations when planning for future transport infrastructure. The research in Chapter 6 has taken a spatial data-driven modelling approach for understanding why potential property value increases were attributed through the provision of new transport infrastructure. This chapter introduces a prototype toolkit known as the Rapid Analytics Interactive Scenario Explorer (RAISE), which applied hedonic price modelling (HPM) to quantify the impact of transport infrastructure on residential property prices. The RAISE toolkit has been initially developed and evaluated in Western Sydney, specifically in the city of Parramatta.

The key research question explored in this chapter was: *Can a rapid analytics decision-support platform be created to support land-use planners' exploration of value uplift scenarios in real-time?* To address this question an early prototype was built using ESRI's ArcGIS and the CommunityViz software extension. The RAISE prototype toolkit has been built to enable city-planners and policy-makers to 'drag and drop' new metro stations, and then the tool automatically calculates the property value increase attributable to the new metro stations to land parcels within a 400m radius, within a 400–800m radius and beyond a 800m radius.

This research and development resulted in an innovative planning support tool which can be used to calculate residential property increase due to new metro infrastructure (value uplift) which are underpinned through hedonic regression-based price models. The RAISE tool kit enables city-planners and others to explore a number of *What if?* scenarios; for example, what if a decision is made to put three new stations along a new metro line, or what if a decision is made to put in two new stations to extend an existing line? This research has demonstrated that such a GIS planning support system can be developed which can provide valuable insights into the likely increases in property prices as driven by new transport infrastructure projects.

Commercial office property and spatial analysis (Chapter 6)

This chapter examined office buildings in the central business district (CBD) of a global city (Melbourne) with the focus placed on spatial characteristics associated with the building's location (Liang, Reed and Crabb 2018). This research is innovative, as it is one of few examinations of the spatial dependency issue in the office building transactional market. This gap in knowledge has been under-researched, although it indexes in other asset classes and land uses, e.g. sharemarket, housing market index, housing construction index. It examined multiple spatial characteristics and how they can assist the construction of a price index for the transactions of entire office buildings and the land component. The construction of a reliable office property index based on locational characteristics would assist analysts and investors to better improve efficiencies in the real estate market. The research question was: *To what extent does spatial dependency exist in the transactional market for office buildings?*

To address the research question a case study was based on examining all transactions of office property buildings between 2000 and 2015, which equated to 289 transactions. Then the methodology developed a transactional price index for the office market to evaluate to what extent incorporating spatial dependency issues into the index could potentially improve the level of accuracy. The analysis commenced by constructed a hedonic price model to determine the relationship between each attribute, including locational attributes, and the price of the combined office building and land.

The core finding was confirmation that spatial dependency existed in the transactional market of office buildings in the case study. In addition, the index based on the hedonic price model showed high levels of volatility, especially during 2007–09, which included the effect of the 2007 global financial crisis. There is strong evidence that including the spatial weights matrix to control for the spatial dependency improves office property transaction indexes. This analysis highlighted the potential contribution of GIS to a better understanding of locational characteristics affecting the value of whole office buildings; this confirms the market is factoring in these considerations in their market evaluations, where GIS assists the assembly of an index to replicate this scenario.

An agent-based model for high-density urban redevelopment under varied market and planning contexts (Chapter 7)

This research conducted in this book presented an agent-based model which has been developed and tested in the context of exploring future redevelopment scenarios of neighbourhood design. In agent-based modelling, the agents, which in this instance are the land parcels, are the foundation spatial unit underpinning the modelling process. The land parcels act as autonomous agents which interact with their environment and each other through behavioural rules. In this research the land parcels and their changing state respond to behavioural rules including suitability, economic feasibility, control, compliance and replacement. The GAMA agent-based modelling platform has been used to construct and test the models.

The key research questions which this chapter addressed are: (i) *What insights can the agent-based model of urban redevelopment provide to users?* (ii) *Can the model be used to investigate potential future scenarios based on neighbourhood design proposals?* and (iii) *What are the main benefits and limitations of agent-based modelling for urban redevelopment related research?* A case study approach developed and tested the agent-based model using the Kensington to Kingsford neighbourhood situated in Randwick City Council, eastern suburbs of Sydney. Four scenarios were developed to test the model including (i) business as usual, (ii) urban redevelopment as determined through an urban design proposal competition, (iii) urban redevelopment combining the urban design proposal with an increase in floor space ratio, (iv) urban development with an automated maximum height revision.

In undertaking this research and testing the agent-based model using a real world case study, this demonstrated the strengths and limitations of the model. The strengths of the model include (i) its ability to easily handle geographic data, (ii) its ability to generate meaningful insights of the combined effects of land market and planning frameworks on the urban redevelopment process, and (iii) the potential of the model to 'plug and play' with urban design proposals to evaluate their potential based on economic feasibility. The model considered two key drivers: land developer goals and planning frameworks. The model does not currently include other key agents in the land development process such as property owners, real estate agents, property buyers, banks and local residents. Future work will extend the model to include these additional key actors in the land redevelopment process.

Architecture, GIS and mapping (Chapter 8)

This chapter focused on the potential contribution from GIS to the architecture discipline (Charles and Reed 2018). From a research perspective there is little information regarding how methodologies such as GIS and mapping can be utilised to provide better insights into architecture. With the focus placed on urban cities, it was argued that the nature of contemporary cities moved to a fragmented and dispersed state due to influencing factors. Accordingly the research question was: *To what extent can GIS and mapping enhance the application of architecture in an urban context?*

The research adopted a conceptual framework based on the following three approaches designed to examine the growth of and change in urban systems: 'observation', 'simulation' and 'speculation'. It was also accepted that most urban cities were perceived as slow-moving entities, therefore examining change over time necessitated alternative viewpoints for varying spans of time. The application of GIS and mapping were potentially able to provide invaluable insights to these viewpoints and deal with associated complexities of non-spatial systems, including the economy, law and politics. To undertake this analysis there were multiple case study examples of real estate in a central business district in an international city. The initial research emphasis was placed on the identification of voids through observation and addressing challenges associated with the voids. This involves assembling and curating drawings, photographs and notes that examined qualitative aspects of voids in the present and also their transitional nature over time; in this role GIS acts as a catalogue system where relevant fields can be entered. Relying upon multiple case study examples the three observational types were: (i) historical formation and growth, (ii) existing use and (iii) potential future use.

The major findings were that all three approaches, namely observation, simulation and speculation, confirmed that architecture can be enhanced by the use of GIS when considering the long-term effect on the built environment incorporating the property and real estate market. Most importantly, the three approaches highlighted different aspects of voids relating to different time periods and locations/areas. Following the contribution of GIS towards an improved understanding of architecture, there were three additional outcomes. First, *observation* provided insights into how voids bind relationships between seemingly disconnected elements and provide useful qualitative frameworks for the spatial qualities of voids. Secondly, *simulation* was useful over longer periods of time. Finally, *speculation* relied on bringing observations and simulations together in a narrative framework where GIS assisted to provide potential future scenarios, e.g. 'what if' scenarios (Pettit 2005).

3D and virtual reality for supporting redevelopment assessment (Chapter 9)

With the rapid growth of many cities, an increasing number of development approvals are being submitted to local councils to process. The research in this chapter explored the use of three-dimensional (3D) modelling and virtual reality to support city-planners when undertaking development approval assessments. Four 3D building models have been constructed based on different levels of detail (LOD). Using these models two studies have been undertaken. First, focus group interviews were conducted involving three local councils across Sydney. Secondly, the perceived utility of 3D models constructed using different LODs and displayed in a virtual reality environment was evaluated through an experiment involving 48 professional planners.

This research addressed two questions. First, *what are the benefits of utilising 3D modelling in local government sectors for DA assessments?* Secondly, *which LODs in a city model can assist planners in assessing DAs?* The first question was addressed through focus group interviews involving seven planning practitioners. The second question was addressed through setting up a digital virtual environment and evaluating the perception of 48 professional planners. The results from the focus group interviews found the current state of practice for assessing development approvals in the three participating local councils involved the use of standard 2D GIS mapping tools with no current utilisation of 3D modelling and visualisation functionality. However, responses from the focus groups indicated they could see potential for 3D modelling to support the assessment process. Some of the benefits discussed in using 3D modelling included providing a better understanding of the spatial dimensions of buildings, arrangement of building, scale, measuring room sizes and to check solar access in relation to the submitted DA. Participants also responded that they perceived value in visualising the entire municipality in 3D to further evaluate the context of the proposed development with respect to building heights and the likely impact on the neighbourhood.

In support, the perceived utility experiment, Unreal Engine 4 developed by computer game developer Epic Games, was used. Four 3D virtual models where imported and customised to represent LOD1, LOD2, LOD3 and LOD3+vegetation. These four 3D models were then visualised and shown to participants in the experiment using the Oculus Rift Virtual Reality (VR) headset. Participants were required to navigate through the virtual environment during the LOD experiment. Results from the experiment showed that all of the four 3D models were perceived as useful, however, LOD3+vegetation was the most useful. This finding indicated that including vegetation information in the development approval assessment can assist in realising the scale of the proposed development and can be used to complete several additional assessments including view shed and visibility analysis. However, as noted by participants, current challenges which remain in the implementation of 3D modelling to support development assessment and other council-related applications include training and upskilling of council staff, the cost of implementing VR technology and VR motion sickness. As 3D modelling and VR technologies become more ubiquitous it would seem that such challenges may be possible to address in the not too distant future.

Challenges in the real estate and GIS disciplines

Drawing upon the research presented in this book it is possible to identify the major challenges facing this field of research.

Data limitations

One of the largest barriers for research into property and real estate-GIS from a global perspective is the availability of data. Whilst this has become less of a problem in recent years, there remain substantial barriers when seeking to obtain detailed information either (i) for a relatively small geographical area, (ii) at small intervals over an extended time period or (iii) both. For example, some of the case studies in this text examined Census data which was taken at five-year intervals, the shortest interval possible and substantially less than the ten-year intervals in some countries. The cost of collecting and organising detailed data for an analysis is often cost-prohibitive due to the time required to undertake these tasks. This is further complicated by the privacy restrictions in many jurisdictions; for example, in these regions it is essential that collected information must be grouped together so individual households cannot be identified.

No dedicated combined real estate-GIS journals to disseminate research

There are no existing journals with a scope surrounding research into combined real estate and GIS. However each discipline has its own research journals. Property and real estate have a limited quantity of journals which are categorised in either the built environment or finance disciplines. GIS is usually associated with journals in the spatial sciences. Note that both disciplines are generally regarded as relatively small in comparison to the established traditional disciplines. In direct contrast, each of the established disciplines is fully supported by a relatively broad series of research journals; furthermore these journals promote their high rankings which are highly regarded by universities, primarily due to the large numbers of researchers in that discipline. This has the effect of ensuring there is a large cohort of researchers in a particular high-profile area, attracting new researchers and high journal rankings which are increasingly sought-after property for university employment retention and promotion. The challenge for researchers in the real estate-GIS combined discipline is to establish high-ranking or perceived high-quality journals in a relatively short period of time.

Relatively low discipline profiles

It can be argued the real estate and GIS disciplines as research fields have relatively low profiles in both universities and in industry. This profile is arguably extremely low for the combined real estate-GIS field. This scenario is regardless of the application and usefulness of research into real estate, GIS or both. The effect is that researchers in these emerging disciplines are competing with established disciplines for profile, which in turn also affects the potential to receive research grants and undertake further research. The challenge is to identify a means of increasing the profile of research into real estate and GIS whilst emphasising the 'real life' contribution of the research outcomes to society. The barriers are the perception of real estate and GIS in broader research institutions and the support needed from an infrastructure perspective.

Attracting new researchers into the real estate-GIS discipline

Few researchers, if any, have identified and committed to a research career in real estate-GIS whilst at school or prior to commencing university. Due to the relatively low profile of these disciplines in comparison to the established traditional disciplines, researchers tend to come from other related disciplines and evolve towards research into real estate, GIS or both. To overcome this barrier there is a requirement to raise the profile of the disciplines and promote the 'real world' relevance of the research outputs. This may involve advising industry of the added value of research into real estate-GIS. For example, real estate is accepted as the largest single asset class, yet few real estate companies enlist the services of a dedicated GIS expert. This may be partly due to the lack of knowledge about GIS and how they can contribute to the outputs of the real estate organisation.

Limited interaction between real estate and GIScience (GIS) disciplines

Due to the location of each discipline in a research institution such as a university, there remains limited opportunity to specialise in research in this area. This can also be observed in the conference proceedings of a real estate or GIS conference with little, if any, combined research published. This is surprising considering the strong synergy between the two disciplines as highlighted in this book. There are clear opportunities, however, as described here, to undertake valuable research in this field, although this may require seeking research opportunities which were not previously clearly apparent.

Future areas for research

The research examined in this book has identified core areas in the real estate-GIS fields which require further analysis. These areas are listed below.

Demography

The linkages between household data and the real estate market are very strong, as confirmed in the research findings. This has been highlighted with varying land uses in different sections of the book, including housing and office land uses. Overall there is a high potential for further research into the locational characteristics of households and their decisions about their geographical tenure. It is accepted that household decisions also have an indirect effect of the operation of economies at the local, regional and country levels due to their choices about where to live, household characteristics and relocation decisions.

Extended time frames

Although the traditional emphasis for most studies is to undertake a static analysis at one point in time, stakeholders linked to the property and real estate-GIS fields are keen to access information about short-term and long-term forecasts. These scenarios are often somewhat based on historical trends which can be evaluated via an analysis of time series data, e.g. five-yearly Census data. In addition there is a need to undertake 'what-if' scenario planning to assist governments to plan ahead; for example, to enable sufficient land development to accommodate population growth. There is need to undertake research over an extended time series which will facilitate an analysis of historical change over time and also permit forward-planning and forecasts.

Multi-dimensional analysis

There have been advancements in GIS which have yet to be fully applied to the real estate sector. To date the primary application for mapping has been linked with two-dimensional (2D) spatial analysis associated with the location of property. However, the development of three-dimensional (3D) research and applications such as 'fly-bys' have substantial potential to assist the understanding of the operation of real estate markets. For example, many planning applications and development applications are submitted in 2D format and do not allow stakeholders to fully evaluate the potential implications of approving the applications (Reed and Sims 2014). There are many examples in urban landscapes of approved developments which were unable to be fully considered as a completed project. It is recommended that future research is undertaken into the application of 3D analysis and its contribution to understanding the operation of the property and real estate sector.

Big data and future developments

It is accepted that changes in GIS occur more rapidly in comparison to the property and real estate sector, therefore both fields need to work closely together to ensure the transfer of information. For example, the emphasis on 'big data' in GIS is an exciting advancement although it has received relatively little attention in property and real estate research. This statement is designed to encourage the property and real estate sector to investigate big data and determine how the application can assist real estate stakeholders to improve their understanding of how the real estate market operates. This is also a catalyst for researchers in both the real estate and GIS fields to work more closely together and identify research areas where there are likely to be future advances.

Summary

This book has provided various innovative research and insights into a range of different real estate property contexts and the contribution of GIS and mapping. This is a unique perspective of this emerging discipline which has become, and will continue to become, increasingly important in this technological age. Most importantly the book has achieved its goals of providing a deeper understanding and the effective tools to undertake meaningful research and advanced knowledge into property and real estate-GIS. An underlying objective was to provide a cross-section of studies which were linked to the common theme of the book, while each chapter was also a standalone research output. The diverse and multidisciplinary nature of property real estate and GIS provided additional challenges related to each research chapter maintaining a balance between each field.

The future for the property and real estate and GIS fields has been confirmed as very positive due to their high relevance. The rapid pace of technological change places substantial pressure on the property and real estate industry to embrace GIS as a means of increasing efficiency, rather than placing insufficient attention on the locational attributes of a property. The property and real estate sector, like many, is under immense pressure to identify operating efficiencies, so it is essential that research into property and real estate-GIS provides a realistic and tangible benefit. This objective was evident throughout the chapters in this book, where the findings and implications for each study were highly relevant to industry, society and the economy. In isolation, research into either property and real estate or GIS provided a limited application. However, the synergy between the two fields has been successfully demonstrated here and it is envisaged this will continue to strengthen over time due to the sustained interest in these areas.

References

Dolan, E.L., Elliott, S.L., Henderson, C., Curran-Everett, D., St. John, K. and Ortiz, P.A. (2017). Evaluating discipline-based education research for promotion and tenure. *Innovative Higher Education*, 43(1), 31–39.

Geertman, S., Allan, A., Pettit, C. and Stillwell, J. (2017). Introduction to 'Planning Support Science for Smarter Urban Futures'. In *Planning Support Science for Smarter Urban Futures*, 1–19. Springer International Publishing.

Jylha, T. and Junnila, S. (2014). The state of value creation in the real estate sector – lesson from lean thinking. *Property Management*, 32 (1), 28–47.

Liang, J., Reed, R.G. and Crabb, T. (2018). The contribution of spatial dependency to office building price indexes: A Melbourne case study. In *Journal of Property Investment and Finance*, 6(3); www.emeraldinsight.com/doi/abs/10.1108/JPIF-03-2017-0021?journalCode=jpif.

Pettit, C. (2005) Use of a collaborative GIS-based planning support system to assist in formulating a sustainable development scenario for Hervey Bay, Australia. *Environment and Planning B: Planning and Design*, 32(4), 523–545.

Pettit, C.J., Tanton, R. and Hunter, J. (2017) An online platform for conducting spatial-statistical analyses of national census data. *Computer, Environment and Urban Systems*, 63, 68–79.

Randolph, B. and Tice, A. (2013) Who lives in higher density housing? A study of spatially discontinuous housing sub-markets in Sydney and Melbourne, *Urban Studies*, 50(3), 2661–2681.

Reed, R.G. and Sims, S. (2014) *Property Development*, 6th edition. Routledge, Abingdon.

APPENDIX A

Glossary of terms

Agent-based modelling (ABM)

A type of computational modelling where the behaviour and interactions of agents are simulated. Agents may represent an individual or a group and can be assigned a set of operational rules that determine how they interact. ABMs can be used to investigate the collective effect of these individual interactions on a system.

Area

Used to describe the surface extent of a unit such as a site, building, suburb or city.

Block

Refers to an allotment of land.

Buffer analysis

A spatial analysis tool commonly used to characterise the locus of incident points relative to a given central location. In Esri ArcGIS, the buffer analysis tool draws concentric zones of specified distances around an area or point of interest. The proximity or dilation of incident points from the area of interest can then be described based on the zone they fall on.

Building area (gross)

The total area of a building measured from the normal outside face of any enclosing walls.

Central business district (CBD)

The main shopping or business area of a town or city, which is usually located in the geographical middle.

Choropleth – heat map

A thematic map that uses colours, shades or patterns to communicate the aggregate values measured for a statistical variable within each unit or geographical area in the map.

CityGML

An international standardised data model for storing and exchanging 3D landscape and city models.

Conforming use

Where the use of a property meets or conforms with town planning requirements.

Discounted cash flow (DCF)

A method of analysing investments where annual cash flows are discounted and accumulated together to arrive at their present value (PV).

Disruptive technology

An innovation which causes a structural change in the way a business, industry or a market operates, or that creates an entirely new industry or market. Disruptive technologies may arise from new inventions or from innovative ways of combining and applying existing technologies and methods.

Dot density map

A map that shows the distribution of incident points for an observed phenomenon, giving a visual indication of where such points may be relatively clustered or dispersed.

Game engine

A software framework that is used for authoring and developing video games and virtual reality for learning. It is a suite of core software that typically includes functionalities for 3D graphics rendering, collision detection and audio which can be modified to build new game applications. With these packaged functionalities, a game engine reduces the amount of underlying programming that game developers would otherwise need to do.

Geographical information science – GIScience

An academic discipline concerned with capturing, formatting and structuring, analysing and representing geographical information. It is an intersection of geography, computer and data science, statistics, cartography and communication.

Geographical information system (GIS)

A system (often referring solely to a computer system, but it could be extended to include hardware, software, human resources and organisational processes) that integrates the management (can include capture, storage, formatting, indexing, syndication) and analysis of data to produce and communicate geographical information.

Hedonic price model

A method for quantifying the economic value of a good, whereby internal and external attributes that characterise the good are weighted and factored into a regression analysis to calculate the price.

Highest and best use

The land use must pass four tests, being (a) physically possible, (b) legally permissible, (c) economically feasible and (d) resulting in maximum productivity.

Improvements

Buildings or alterations to land, e.g. a structural improvement, clearing of a site.

Index

An indicator which represents the value of the properties it contains.

Land use

Main use of the property, e.g. residential, retail, office, etc. (see Appendix B).

Level of detail (LOD)

The degree of complexity rendered in a 3D model representation. Application of LOD techniques involves decreasing the complexity of an object's 3D representation relative to the user's viewpoint to increase the efficiency of 3D rendering.

Location quotient

A ratio used to quantify the concentration of a phenomenon within a local area relative to the phenomenon's occurrence over a wider geographical region. See Research methodology in Chapter 4.

Lot

A portion of any land (except government land) which can be sold separately.

Market rental value

The estimated amount for which an asset should rent, as at the relevant date, between a willing lessor and a willing lessee in an arm's-length transaction, wherein the parties had each acted knowledgeably, prudently and without compulsion, and having regard to the usual terms and conditions for leases of similar assets.

Market value

Market value is generally accepted as the estimated amount for which an asset should exchange on the date of valuation between a willing buyer and a willing seller in an arm's-length transaction after proper marketing, wherein the parties had each acted knowledgeably, prudently and without compulsion.

Moran's I

An autocorrelation tool used to determine whether the distribution of values measured for an observed variable over a set of spatial features is random, dispersed or clustered. Anselin Local Moran's I is a spatial autocorrelation tool that identifies (i) statistically significant clusters of neighbouring features with similar high or low values, and (ii) outlier features that have statistically higher or lower values compared to their neighbours. See Research methodology in Chapter 4.

Non-conforming use

A use which is prohibited due to rezoning.

Permissible use

A use of the land which is allowed.

Planning support system (PSS)

A system that may integrate datasets, user inputs, computational modelling and visualisation capabilities which planners can use to test their hypotheses and estimate potential outcomes of their activity.

Property

This usually refers to the private rights of ownership. It is often used interchangeably with reference to real estate.

Real estate

Land and improvements which are associated with the land. It is often used interchangeably with reference to property.

Regression

A type of statistical modelling that is used to predict the value of a dependent variable (e.g. market price) based on a set of attributes (independent variables) that positively or negatively affect the value of the dependent variable.

Retail

Selling services, goods or materials direct to the public.

Sharing economy

An alternative to the business-to-consumer mode of exchanging goods and services, where goods and services are directly supplied and consumed by peers and transactions are facilitated by online-based platforms.

Site

Land which is usually not improved.

Spatial statistics

A set of formal techniques used to analyse the geographic, topological and geometric characteristics of an entity or phenomena over space and time.

Tenant mix

Different retailers and services that occupy a shopping centre.

Value

The quantity of something obtained in exchange for another. The value of an object is usually equal to the present value of the future benefits.

Value capture

A public infrastructure funding mechanism wherein the cost of delivering new infrastructure is recovered from private landowners who benefit from the value uplift generated by the new infrastructure.

Value uplift

The increase in land value that results from the capitilisation of the land's proximity or access to new infrastructure.

Virtual environment

A computer-generated environment where physical elements from a real or hypothetical world setting are simulated. See Virtual reality.

Virtual reality (VR)

A computer technology that enables users to explore physical elements (e.g. sight and sound) of a simulated environment. More sophisticated VR systems enable users to interact with items in the virtual environment and experience feedback from the interaction through the transmission of vibrations to users' haptic devices (e.g. game controller).

Zoning

The legally allowable uses of land as set out by a planning authority.

APPENDIX B

Property classification codes

This appendix lists an example of Valuation Property Classification Codes (using Australia as a case study) and groups them into categories based on a specific land use.

This information was supplied by the Valuer-General Victoria at the Department of Environment, Land, Water and Planning.

Code	Residential	Description
10	**Residential Use Development Land**	
100	Vacant Residential Dwelling Site/ Surveyed Lot	Vacant land suitable for the erection of a detached or semi-detached dwelling.
101	Vacant Residential Development Site	Vacant land with a permit approved or capable of being developed for high-density residential purposes.
102	Vacant In globo Residential Subdivisional Land	Vacant land zoned for future residential subdivision.
102.2	*Subdivisional Land (Multi Lot)*	
102.3	*Subdivisional Land (In globo / Potential)*	
103	Vacant Residential Rural/Rural Lifestyle	Residential Rural/Rural Lifestyle in a rural, semi-rural or bushland setting that has not been developed.
104	Vacant Residential Rural/Rural Lifestyle (with permit refused)	Residential Rural/Rural Lifestyle in a rural, semi-rural or bushland setting that has not been developed where a permit for a dwelling has been rejected.
109	Residential Airspace	Airspace capable of being developed for residential purposes, usually above a rooftop, roadway or railway.

11	*Single Residential Accommodation*	
110	Detached Dwelling	Freestanding dwelling on residential land.
110.2	*Detached Dwelling (new)*	
110.3	*Detached Dwelling (existing)*	
110.4	Detached Dwelling *Non-Conforming Use-Commercial*	Detached dwelling used for residential purposes on commercial land.
110.5	Detached Dwelling *Non-Conforming Use-Industrial*	Detached dwelling used for residential purposes on industrial land.
111	Separate Dwelling and Curtilage	Freestanding dwelling on defined curtilage, being part of a larger holding of varying use, e.g. caretaker's dwelling on industrial site, second dwelling on farm.
112	Semi-Detached/Terrace/Row House	Attached or semi-detached dwelling that does not share common land.
112.2	*Semi-detached*	
112.3	*Terrace*	
112.4	*Row House*	
112.5	*Half Pair or Duplex*	
113	Granny Flat/Studio	Dwelling either detached or adjoined to a principal dwelling that is self-contained, connected to services but not subdivided in the present form.
114	Dwelling and Dependant Unit	Dwelling and dependant unit not permitted to be separately occupied either detached or adjoined, each self-contained, connected to services but not subdivided in the present form.
115	Shack/Hut/Donga	Basic structure with limited services and amenities, providing basic, short-term accommodation.
116	Cabin/Accommodation (rental/ leased individual residential site)	A dwelling subject to a site agreement within a larger complex, e.g. caravan park or lifestyle village. The basis of valuation may vary according to legislation.
117	Residential Rural/Rural Lifestyle	A single residential dwelling on land in a rural, semi-rural or bushland setting. Primary production uses and associated improvements are secondary to the residential use.
118	Residential Land (with buildings that add no value)	Residential land on which the benefit of works (structures erected) upon the land is exhausted.
12	*Multiple Occupation (within residential development)*	
120	Single Strata Unit/Villa Unit/ Townhouse	Freestanding and unattached unit with ground level footprint.
120.2	*Single Strata Unit*	
120.3	*Villa Unit*	
120.4	*Townhouse*	
120.5	*OYO Subdivided Dwelling*	
120.6	*OYO Subdivided Unit*	
120.7	*OYO Unit*	

121	Conjoined Strata Unit/Townhouse	Unit/Townhouse with common walls/party walls with ground level footprint.
121.2	*Half Pair or Duplex*	
121.3	*Conjoined Strata Unit*	
121.4	*Townhouse*	
123	Residential Company Share Unit (ground level)	Unit with a ground level footprint where the land is defined by shares in a complex, which gives right to occupy a particular unit. Fully serviced and equipped for long-term residential occupation.
123.2	*OYO Company Share Unit*	
123.3	*OYO Stratum Flat*	
123.4	*OYO Stratum Unit*	
123.5	*OYO Company Share Flat*	
123.6	*OYO Cluster Unit*	
124	Residential Company Share Unit (within multi-storey development)	Unit which forms part of a multi-storey development where the land is defined by shares in a complex which gives rights to occupy a particular unit. Fully serviced and equipped for long-term residential occupation.
125	Strata unit or flat	Unit or flat that forms part of a multi-storey development that has been subdivided.
125.2	*Strata Dwelling*	
125.3	*OYO Subdivided Flat*	
125.4	*OYO Strata Flat*	
126	Individual Car Park	Individual car park associated with residential use.
127	Individual Berth	Individual berth associated with residential use.
128	Individual Flat	Single flat used for residential purposes within a larger property, e.g. caretaker's flat or dwelling above a shop.
129	Common Land associated with a residential development	Designated common space, e.g. driveway, gardens or common parking.
13	*Investment Residential*	
130	Boarding House	Land with a building that is registered to provide long-term single room accommodation with shared facilities.
131	Residential Investment Flats	A flat that forms part of a complex of two or more flats on land not subdivided.
133	Short Term Holiday Accommodation	A residential property used exclusively for short-term holiday accommodation for persons away from their normal place of residence.
135	Dormitory Accommodation/ University Residential College	Residential accommodation for students usually with shared facilities which is controlled or operated by a registered educational provider.
135.2	*Dormitory Accommodation*	
135.3	*University Residential College*	
14	*Retirement/Aged Care Accommodation/Special Accommodation*	
140	Retirement Village Unit	Individual unit with self-contained facilities that forms part of retirement village.

141	Retirement Village Complex	Land on which a retirement village complex, which provides accommodation with shared facilities, amenities and services, is erected.
142	Aged Care Complex	Land developed with a registered care facility that provides residential accommodation and care services for people, particularly the elderly, who can no longer live independently.
142.2	*Aged Care Complex*	
142.3	*Nursing Home*	
143	Special Accommodation	Land with residential accommodation provided by registered providers for people with defined medical, social or special support needs.
144	Disability Housing	Land on which purpose-built housing accommodation for people with disabilities is erected.
15	**Ancillary Buildings**	
150	Miscellaneous Improvements on Residential Land	Residential land, on which ancillary improvements only are erected.
150.2	*Storage Area*	
150.3	*Garage/Outbuilding*	
151	Miscellaneous Improvements on Residential Rural Land	Rural Residential land on which ancillary improvements only are erected.
151.2	*Storage Area*	
151.3	*Garage/Outbuilding*	
151.4	*Site Improvements*	

Code	Commercial	Description
20	*Commercial Use Development Land*	
200	Commercial Development Site	Vacant land with a permit approved or capable of being developed for commercial or mixed use purposes.
201	Vacant In globo Commercial Land	Land which is zoned for future commercial subdivision/development.
202	Commercial Land (with buildings that add no value)	Commercial land on which the benefit of works (structures erected) upon it is exhausted.
209	Commercial Airspace	Airspace capable of being developed for commercial purposes, usually above a rooftop, roadway, railway.
21	**Retail**	
210	Retail Premises (single occupancy)	Land with retail premises used for the sale of goods or services.
210.2	*Bank*	
210.3	*Retail Store/Showroom*	
210.4	*Shop*	
210.5	*Café*	
210.6	*Timber Yard/Trade Supplies*	
210.7	*Display Yard*	

210.8	*Convenience Store / Fast Food*	
210.9	*Plant / Tree Nursery*	
211	Retail Premises (multiple occupancies)	Land with more than one retail premises used for the sale of goods or services, regarded as a complex and not subdivided.
211.1	*Unspecified*	
211.2	*Shop and Dwelling (single occupancy)*	
211.3	*Office and Dwelling (single occupancy)*	
212	Mixed Use Occupation	Land that includes mixed occupancies, including shops and offices regarded as a complex and not subdivided.
212.3	*Office*	
212.4	*Shop*	
212.5	*Studio*	
212.6	*Workroom*	
213	Shopping Centre	Land developed with a significant retail complex comprising a number of unsubdivided retail premises, parking and associated infrastructure.
213.2	*Super Regional*	
213.3	*Major Regional*	
213.4	*Regional*	
213.5	*Sub Regional*	
213.6	*Neighbourhood*	
214	National Company Retail	Land developed with a purpose-built structure and normally occupied by a national company, e.g. supermarket, hardware and home wares.
214.2	*Supermarket*	
214.3	*Department / Discount Department Store*	
214.4	*Bulky Goods*	
215	Fuel Outlet/Garage/Service Station	Land used predominantly for fuel sales (multiple pumps), may include car repair and servicing facilities.
216	Multi-Purpose Fuel Outlet (fuel/ food/groceries)	Land used as a service centre usually including fuel sales, retail, restaurant and takeaway food facilities.
217	Bottle Shop/Licensed Liquor Outlet	Land developed with a purpose-built structure and normally occupied by a national company licensed for the sale of packaged alcohol.
218	Licensed Retail Premises	Retail premises licensed to sell packaged alcohol.
219	Market Stall	A stall within a market used for the sale of goods and services, e.g. stall at Queen Victoria Market.
22	*Office*	
220	Office Premises	Land used as an office for administration, technical, professional or other like business activity.
220.1	*Unspecified*	
220.2	*Office (Converted dwelling)*	
220.3	*Serviced Office*	
220.4	*Strata / Subdivided Office*	

221	Low Rise Office Building	Land developed with a one- to three-level office building and used for administration, technical, professional or other like business activity.
222	Multi-Level Office Building	Land developed with a four+ level office building and used for administration, technical, professional or other like business activity.
222.2	*Medium rise (4 to 50 levels)*	
222.3	*High Rise (50+ levels)*	
223	Special Purpose (built-in technology)	Land developed as a purpose-built facility with a high technology component, e.g. call centre.
23	**Short Term Business and Tourist Accommodation**	
230	Residential Hotel/Motel/Apartment Hotel Complex	Land used to provide accommodation in serviced rooms for persons away from their normal place of residence.
230.2	*Residential Hotel*	
230.3	*Motel*	
230.4	*Apartment Hotel Complex*	
230.5	*Tourist Resort Complex*	
230.6	*Hotel*	
230.7	*Private Hotel*	
231	Residential Hotel/Motel/Apartment Hotel Units	Subdivided units forming part of a single complex operated as a hotel/motel.
232	Serviced Apartments/Holiday Units	Unit/s within a development used to provide short-term accommodation as serviced apartments.
232.2	*Holiday Units*	
232.3	*Serviced Apartments*	
233	Bed and Breakfast	Land developed with short-term accommodation, permitted in serviced rooms for persons away from their normal place of residence.
234	Tourist Park/Caravan Park/Camping Ground	Land registered as a caravan park and developed with cabins, caravan and camping sites, administration/ablution amenities and recreational facilities.
235	Guest Lodge/Back Packers/ Bunkhouse/Hostel	Land providing basic, short-term residential accommodation usually with shared bathroom and self-service catering facilities.
235.2	*Guest Lodge*	
235.3	*Back Packers/ Hostel*	
235.4	*Bunkhouse*	
236	Ski lodge/ Member facility	Land developed with short-term accommodation for members or guests of a ski lodge/club.
237	Recreation Camp	Land developed with accommodation used by persons or groups for holiday or recreational purposes provided by a commercial operator, e.g. student, youth or family groups.

24	*Hospitality*	
240	Pub/Tavern/Hotel/Licensed Club/ Restaurant/Licensed Restaurant/ Nightclub	Land licensed to sell liquor but not permitted to provide gaming facilities. May provide meals, limited accommodation and/or entertainment.
240.2	*Pub*	
240.3	*Tavern*	
240.4	*Hotel*	
240.5	*Licensed Club*	
240.6	*Restaurant*	
240.7	*Licensed Restaurant*	
240.8	*Nightclub/Cabaret*	
240.9	*Reception/Function Centre*	
241	Hotel–Gaming	Land licensed to sell liquor and permitted to provide gaming facilities. May provide meals, limited accommodation and/or entertainment.
242	Club–Gaming – stand alone	Land permitted to provide gaming facilities associated with a special purpose organisation, e.g. ethnic club, RSL. Access is normally subject to entry conditions.
243	Member Club Facility	Land upon which the use of the facilities is restricted by membership requirements. Entry is not available to non-members. May contain any combination of liquor sales, meals and limited accommodation, e.g. RACV member club, The Australian Club.
244	Casino	Land with special operating permit for a large gaming facility.
245	National Company Restaurant	Land occupied by a national company and used as a fast food outlet, e.g. KFC, McDonald's.
246	Kiosk	Land developed with a small retail facility commonly found in public areas, e.g. parks, transport hubs.
247	Conference/Convention centre	Land developed with purpose-built facilities used for conference or convention centre purposes.
25	*Entertainment – Cinema, Live Theatre and Amusements (non-sporting)*	
250	Live Entertainment – Major Multi- Purpose Complex	Land developed with a large purpose-built venue used for a wide variety of live entertainment, e.g. Melbourne Arts Centre.
251	Cinema Complex	Land developed as a cinema complex incorporating theatres, either standalone or within a larger complex.
252	Playhouse/Traditional Theatre	Land developed as a theatre, either standalone or within a larger complex.
253	Drive-In	Land with an outdoor movie theatre with drive-in parking facilities.
26	*Tourism Facilities/Infrastructure*	
260	Large Theme Attraction/Park	Land developed as a high-profile theme park with attractions, e.g. Sovereign Hill, Ballarat.

261	Amusement Park	Land developed as purpose-built amusement park with limited rides and attractions, e.g. Luna Park, Melbourne.
262	Major Infrastructure Attractions (often associated with a major historic or feature natural location).	Land associated with a major tourist attraction destination, e.g. Penguin Parade at Phillip Island, Otway Fly Treetop Walk.
263	Tourism Infrastructure – Local Attractions	Land associated with a local tourist attraction, e.g. cable cars, water slides, chair lifts, tourist railways.

27   *Personal Services*

270	Health Surgery	Land used by a health practitioner in a standalone practice.
271	Health Clinic	Land used as consulting suites by health practitioners within an unsubdivided complex, e.g. doctor, chiropractor, dentist, radiologist.
271.2	*Diagnostic Centre / X-Ray*	
271.3	*Medical Centre / Surgery*	
271.4	*Dental Clinic*	
271.5	*Super Clinic*	
272	Brothel	Land permitted to be used for the business of providing prostitution services.
273	Crematorium/Funeral Services	Land that is purpose-built for undertaking funeral services.
274	Automatic Teller Machine	An ATM facility that is not within or attached to banking premises. Can be standalone or separately occupied.
275	Veterinary Clinic	Land used by a veterinary practitioner to treat animals. It may include keeping animals on the premises for treatment or adoption.

28   *Vehicle Car Parking, Washing and Sales*

280	Ground Level Parking	Land used for ground level parking.
281	Multi-Storey Car Park	Land developed as a multi-storey car parking facility.
282	Individual Car Park Site	A subdivided car park within a commercial property. Can be leased individually or as part of a single complex by a car park operator.
282.2	*Car park – Under Cover*	
282.3	*Car park – Open Air*	
283	Car Wash	Land developed as a purpose-built car wash facility. Can be standalone or part of a larger property.
284	Vehicle Sales Centre	Land used for the preparation and display of new or second-hand vehicles for sale.
285	Vehicle Rental Centre	Land used for the preparation, storage and display of vehicles available for hire.

29   *Advertising or Public Information Screens*

| 290 | Advertising Sign | Land upon which an advertising sign is erected; may be standalone or form part of a larger property. |

290.1 *Unspecified*
290.2 *Self-Standing Pole*
290.3 *Bridge Fixed*
290.4 *Roof Mounted*
290.5 *Wall Fixed*

293	Electronic Stadium/Street TV Relay Screen/Scoreboard	Land upon which electronic screen or scoreboard used for the display of live media and information is erected. May be standalone or form part of a larger property.

Code	Industrial	Description
30	**Industrial Use Development Land**	
300	Industrial Development Site	Vacant land with a permit approved or capable of being developed for industrial use.
301	Vacant Industrial In globo Land	Land which is zoned for future industrial subdivision/development.
302	Industrial Airspace	Airspace capable of being developed for industrial purposes, usually above a rooftop, roadway, railway.
303	Industrial Land (with buildings which add no value)	Industrial land on which the benefit of works (structures erected) upon it is exhausted.
31	**Manufacturing**	
310	General Purpose Factory	Land used for manufacturing, assembly or repairs. May have specialised/purpose-built structures.
310.2	*Factory Unit*	
310.3	*Factory*	
310.4	*Garage/Motor Vehicle Repairs*	
310.5	*Office/Factory*	
310.6	*Workshop*	

Code	Industrial	Description
311	Food Processing Factory	Land developed with purpose-built food processing facilities, e.g. cannery, milk production plant.
311.2	*Processing Plant*	
311.3	*Dairy*	
312	Major Industrial Complex – Special Purpose Improvements	Land developed with purpose-built facilities for large scale industrial use, e.g. car plant, paper mills.
32	**Warehouse/Distribution/Storage**	
320	General Purpose Warehouse	Land used for the storage of goods.
320.2	*Warehouse*	
320.3	*Warehouse/Office*	
320.4	*Warehouse/Factory*	
320.5	*Warehouse/Showroom*	

320.6	*Depot*	
320.7	*Store*	
321	Open Area Storage	Land with extensive hardstand area used for the storage of goods and equipment.
321.2	*Hardstand / Storage Yard*	
321.3	*Wrecking Yard*	
321.4	*Concrete Batching Plant*	
321.5	*Container storage*	
322	Bulk Grain Storage (structures)	Land developed with silos used for the storage of grain.
323	Bulk Grain Storage (earthen walls and flooring – pit bunker)	Land developed with bunkers used for the storage of grain.
324	Bulk Liquid Storage Fuel Depot/ Tank Farm	Land developed with tanks for the storage and distribution of bulk liquids, e.g. tank farms, fuel depot.
325	Coolstore/Coldstore	Land with a purpose-built structure used for the cold storage of perishable products.
326	Works Depot	Land developed as a works depot used in conjunction with infrastructure maintenance, e.g. municipal depot.
33	**Noxious / Offensive / Dangerous Industry**	
330	Tannery/Skins Depot and Drying	Land developed for the tanning of skins and hides.
331	Abattoirs	Land developed with purpose-built structures used for the holding and slaughter of stock and the preparation of meat for the wholesale market.
332	Stock sales yards	Land developed with purpose-built structures used for the yarding and selling of stock.
333	Rendering Plant	Land developed with purpose-built structures used for the extraction of lard, tallow and oil from animal parts.
334	Oil Refinery	Land developed with purpose-built structures used in the refinement and storage of petroleum products.
335	Petro-chemical Manufacturing	Land developed with purpose-built structures used in the production of chemical-based products from petroleum.
336	Sawmill	Land developed with purpose-built structures used for the milling and curing of timber.

Code	*Extractive Industries*	*Description*
40	**Extractive industry site with permit or reserve not in use**	
400	Sand	Land permitted to be used for the extraction of building/manufacture materials (silica).
401	Gravel/Stone	Land permitted to be used for the extraction of materials used for road works/construction.

402	Manufacturing Materials	Land permitted to be used for the extraction of materials used for manufacturing, such as clay (paper and pottery), limestone, dolomite (fertiliser) and cement/gypsum (cement).
403	Soil	Land permitted to be used for the extraction of soil.
404	Coal	Land permitted to be used for the extraction of coal.
405	Minerals/Ores	Land permitted to be used for the extraction of various types of minerals and ore.
406	Precious Metals	Land permitted to be used for the extraction of precious metals, e.g. gold, silver.
407	Uranium	Land permitted to be used for the extraction/storage of uranium.
408	Quarry/Mine (open cut) – Exhausted (dry)	Land formerly used for extractive industries whose materials have been exhausted.
409	Quarry/Mine (open cut) – Exhausted (wet)	Land formerly used for extractive industries whose materials have been exhausted and is now inundated with water.
41	***Quarry (in use)***	
410	Sand	Land from which sand is being extracted by a licensed operator.
411	Gravel/Stone	Land from which stone and gravel are being extracted by a licensed operator.
412	Manufacturing Materials	Land from which manufacturing materials, such as clay (paper and pottery), limestone, dolomite (fertiliser) and cement/gypsum (cement), are being extracted by a licensed operator.
413	Soil	Land from which soil is being extracted by a licensed operator.
42	***Mine (open cut)***	
420	Black or Brown Coal	Land from which black or brown coal is being extracted by a licensed operator.
421	Iron Ore	Land from which iron ore is being extracted by a licensed operator.
422	Bauxite	Land from which bauxite is being extracted by a licensed operator.
423	Gold	Land from which gold is being extracted by a licensed operator.
424	Metals (other than gold)	Land from which metals (other than gold) are being extracted by a licensed operator.
425	Precious Stones	Land from which precious stones are being extracted by a licensed operator.
426	Uranium	Land from which uranium is being extracted by a licensed operator.
427	Non Metals (other than Uranium)	Land from which non-metals (other than uranium) are being extracted by a licensed operator.

43	*Mine (deep shaft)*	
430	Non-metals	Land from which non-metals are being extracted by a licensed operator, from deep underground, by way of an inclined or vertical passageway or shaft equipped with lifting machinery.
431	Black Coal	Land from which black coal is being extracted by a licensed operator from deep underground, by way of an inclined or vertical passageway or shaft equipped with lifting machinery.
432	Precious Stones	Land from which precious stones are being extracted by a licensed operator from deep underground, by way of an inclined or vertical passageway or shaft equipped with lifting machinery.
433	Gold	Land from which gold is being extracted by a licensed operator from deep underground, by way of an inclined or vertical passageway or shaft equipped with lifting machinery.
434	Metals (other than gold)	Land from which metals (other than gold) are being extracted by a licensed operator from deep underground, by way of an inclined or vertical passageway or shaft equipped with lifting machinery. Gold may be extracted from ore but it is not the principal mining activity.
439	Closed Mine Shaft	Land containing a decommissioned mine shaft, where above ground structures may remain.

44	*Tailings Dumps*	
440	Tailings Dump (minerals)	Land permitted to be used for the storage/treatment of minerals in tailing dumps and dams.
441	Tailings Dump (non-minerals)	Land permitted to be used for the storage/treatment of non-minerals in tailing dumps and dams.

45	*Well/Bore*	
450	Oil	Land containing a narrow hole drilled or dug into the earth for the production of oil.
451	Gas	Land containing a narrow hole drilled or dug into the earth for the production of natural gas.
452	Water (mineral)	Land containing a narrow hole drilled or dug into the earth for the production of water (mineral).
453	Water (stock and domestic)	Land containing a narrow hole drilled or dug into the earth for the production of water, for use for stock and domestic purposes.
454	Water (irrigation)	Land containing a narrow hole drilled or dug into the earth for the production of water, for use in irrigation.

459	Disused Bore/Well	Land containing a narrow hole drilled into the earth that is decommissioned. Above ground structures may remain.
46	**Salt Pan (evaporative)**	
460	Lake – Salt Extraction	Land containing a lake from which salt is extracted.
461	Man-made Evaporative Basin	Land containing a man-made evaporative basin from which salt is extracted.
47	**Dredging Operations**	
470	Dredging (minerals)	Land on which dredging for the extraction, treatment and restoration of submerged minerals occurs. Usually licensed operations on Crown/State land subject to inundation.
471	Dredging (non-minerals)	Land on which dredging for the extraction, treatment and restoration of submerged materials that are not minerals occurs. Usually licensed operations on Crown/State land subject to inundation.
48	**Other Unspecified**	
480	Extractive less than 2 Metres	Land from which material is extracted, that do not exceed 2m in depth.
481	Operating mine unspecified	Land from which material is extracted, but is not otherwise specified.
482	Vacant Land mining unspecified	Land from which material has been extracted in the past, that is decommissioned, is vacant land and not otherwise specified.

Code	Primary production	Description
50	**Native Vegetation**	
500	Vacant Land – Native Vegetation/ Bushland	Vacant land that is not cleared with native vegetation coverage typical of the district that is not covered by a covenant or other formal agreement to preserve the vegetation.
501	Vacant Land – Native Vegetation/ Bushland with Covenant/ Agreement	Vacant land that is not cleared with native vegetation coverage typical of the district, covered by a covenant or other formal agreement to preserve the vegetation.
51	**Agriculture Cropping**	
510	General Cropping	Land used for the production of broad-acre crops, e.g. grains, oilseeds and cotton.
510.2	*Crop Production – Mixed/Other*	
510.3	*Crop Production – Other Grains/Oil Seeds*	
510.4	*Crop Production – Wheat*	
510.5	*Crop Production – Fodder Crops*	

| 511 | Specialised Cropping | Land used in the production of broad-acre crops that require a permit, licence or specialist infrastructure, e.g. pyrethrum, poppies. |

52   **Livestock Grazing**

520	Domestic Livestock Grazing	Land used for the grazing of domestic livestock.
521	Non-Native Animals	Land used for the grazing of specialist/exotic animals.
522	Native Animals	Land used for the grazing of native animals.
523	Livestock Production – Sheep	Land developed with specialist infrastructure and used for the farming of sheep.
524	Livestock Production – Beef Cattle	Land developed with specialist infrastructure and used for the farming of beef cattle.
525	Livestock Production – Dairy Cattle	Land developed with specialist infrastructure and used for the farming of dairy cattle.

53   **Mixed Farming and Grazing**

530	Mixed farming and grazing	Land used for mixed use farming purpose, e.g. cropping and grazing/livestock production.
530.2	*Mixed farming and grazing with infrastructure*	
530.3	*Mixed farming and grazing without infrastructure*	

54   **Livestock – special purpose fencing, pens, cages, yards or shedding, stables**

540	Cattle Feed Lot	Land developed with specialist infrastructure used for intensive feeding of cattle.
541	Poultry – Open Range	Land used for poultry run as free-range.
542	Poultry (egg production)	Land developed with specialist infrastructure used for egg production.
543	Poultry (broiler production)	Land developed with specialist infrastructure used for broiler production.
544	Horse Stud/Training Facilities/ Stables	Land developed with specialist infrastructure used as a horse stud farm or horse training facility.
544.2	*Horse Stud*	
544.3	*Training Facilities*	
544.4	*Stables*	
545	Piggery	Land developed with specialist infrastructure for use as a piggery.
546	Kennel/Cattery	Land developed with specialist infrastructure for use as a kennel and/or cattery.

55   **Horticulture Fruit and Vegetable Crops**

| 550 | Market Garden – Vegetables | Land used for the planting of vegetable crops. |
| 551 | Orchards, Groves and Plantations | Land used for the planting of trees for the production of fruit and nuts, e.g. olives, stone fruits, tropical fruits, citrus. |

56   **Horticulture – Special Purpose Structural Improvements**

| 561 | Vineyard | Land developed with specialist infrastructure to facilitate the growing of grapes. |

562	Plant/Tree Nursery	Land used for the propagation, growing and storage of plants.
563	Commercial Flower and Plant Growing – (outdoor)	Land used for the propagation, growing and storage of flowers and plants.
564	Glasshouse Plant/Vegetable Production	Land developed with specialist infrastructure for the indoor propagation and growing of plants and plant crops.
57	**Forestry – Commercial Timber Production**	
570	Softwood Plantation	Land used for the growing and harvesting of softwood trees, e.g. Radiata Pine.
571	Hardwood Plantation	Land used for the growing and harvesting of hardwood trees, e.g. Blue Gum.
572	Native Hardwood (standing timber)	Land used for the growing and harvesting of native trees within a revegetated area.
58	**Aquaculture**	
580	Oyster Beds	Land developed with specialist infrastructure used for the cultivation of oysters.
581	Fish Farming – Sea Water Based	Land developed with specialist infrastructure used for sea-water based fish farming.
582	Yabby Farming	Land developed with specialist infrastructure used for yabby farming.
583	Aquaculture Breeding/Research Facilities/ Fish Hatchery	Land upon which special purpose aquaculture breeding and/or research facilities are constructed. Includes the breeding and growing of fish for commercial purposes.

Code	Infrastructure and utilities (industrial)	Description
60	**Vacant**	
600	Vacant Land	Vacant land reserved or capable of being developed for infrastructure purposes.
601	Unspecified – Transport, Storage, Utilities and Communication	Vacant land reserved or capable of being developed for transportation, storage, utilities and communication uses.
61	**Gas or Fuel**	
610	Wells	Land developed with specialist infrastructure used as a gas or fuel well.
611	Production/Refinery	Land developed with specialist infrastructure used for the production/refinery of gas or fuel.
612	Storage	Land developed with specialist infrastructure used for the storage of gas or fuel.
613	Transmission Pipeline (through easements, freehold and public land)	Land developed with specialist infrastructure used for the transmission of gas or fuel including pipelines and pressure control facilities.

614	Distribution/Reticulation Pipelines (through easements, freehold and public land)	Land developed with specialist infrastructure used for the reticulation of gas or fuel for domestic/commercial purposes.
62	*Electricity*	
620	Electricity Power Generators – Fuel Powered (includes brown coal, black coal, natural steam, gas, oil and nuclear)	Land developed with specialist infrastructure used in the generation of fossil-fuelled electricity.
621	Hydroelectricity Generation	Land developed with specialist infrastructure used in the generation of hydroelectricity.
622	Wind Farm Electricity Generation	Land developed with specialist infrastructure used in the generation of wind-powered electricity.
623	Electricity Substation/Terminal	Land developed with specialist infrastructure associated with the reticulation of electricity.
624	Electricity Transmission Lines (through easements, freehold and public land)	Land developed with transmission lines used for electricity transmission.
625	Electricity Distribution/Reticulation Lines (through easements, freehold and public land)	Land developed with specialist infrastructure with a transmission line used for domestic/commercial reticulation.
626	Solar Electricity Generation	Land developed with specialist infrastructure used in the generation of solar electricity.
63	*Waste Disposal, Treatment and Recycling*	
630	Refuse Incinerator	Land developed with specialist infrastructure used for the incineration of refuse.
631	Refuse Transfer Station	Land developed with specialist infrastructure used in the storage and transfer of refuse.
632	Sanitary Land Fill	Land permitted to be used for the disposal of household, commercial, industrial and public waste.
633	Refuse Recycling	Land developed with specialist infrastructure used in the recycling of refuse.
634	Hazardous Materials/Toxic Storage Centre	Land permitted to be used for the storage of hazardous materials and toxic waste.
635	Toxic By-product Storage and Decontamination Site	Land permitted to be used for the storage of mining waste.
636	Sewerage/Stormwater Treatment Plant Site	Land developed with specialist infrastructure used in the treatment of sewerage and stormwater.
637	Sewerage/Stormwater Pump Stations	Land developed with specialist infrastructure used in the pumping of sewerage and stormwater.
638	Sewerage/Stormwater Pipelines (through easements, freehold and public land)	Land developed with pipelines or channels used for domestic sewerage or stormwater reticulation.
638.2	*Public Utility – Drainage*	
638.3	*Public Utility – Sewerage*	
638.4	*Reserve for Drainage or Sewerage Purposes*	

638.5   *Retarding Basin*

64   **Water Supply**

640   Water Catchment Area   Land used for the purpose of water catchment
                                               within a designated water catchment area.

641   Water Catchment Dam/Reservoir   Land developed with specialist infrastructure and
                                               used as a dam, weir, storage basin or reservoir
                                               for water catchment.

642   Water Storage Dam/Reservoir   Land developed with specialist infrastructure and
            (Non-Catchment)   used as a dam, weir, storage basin or reservoir
                                               for water storage.

643   Water Treatment Plant   Land developed with specialist infrastructure
                                               used for the treatment of water, e.g.
                                               desalination plant.

644   Water Storage Tanks, Pressure   Land developed with water storage tanks,
            Control Towers and Pumping   pressure control towers and pumping stations
            Stations.   used for water supply.

645   Major Water Conduits   Land developed with canals, flumes, pipes to
                                               carry water to power stations, treatment plants
                                               and irrigation supply channels used for the
                                               supply of water.

646   Water – Urban Distribution   Land developed with infrastructure for the
            Network (through easements,   domestic reticulation of water.
            freehold and public land)

65   **Transport – Road Systems**
650   Freeways   Land that forms part of a freeway.
651   Main Highways (including national Land that forms part of a main highway.
            routes)
652   Secondary Roads   Land that forms part of a secondary road.
653   Suburban and Rural Roads   Land that forms part of a suburban or rural road.
654   Closed Roads   Land that forms part of a road that is now closed.
655   Reserved Roads   Land reserved for future roads.
656   Bus Maintenance Depot   Land developed for the parking and maintenance
                                               of passenger buses.
657   Bus Interchange Centre/Bus   Land developed as a bus interchange centre/bus
            Terminal   terminal.
658   Designated Bus/Taxi Stops/Stands/ Land developed as a bus/taxi stop. Includes
            Shelters   designated areas, stands and shelters.
659   Weighbridge   Land developed with a weighbridge.

66   **Transport – Rail and Tramway**
            Systems
660   Railway Line in use   Land developed and used as an operating railway
                                               line and associated infrastructure.

661   Railway Switching and Marshalling Land developed and used as a railway switching
            Yards   and marshalling yard.
662   Railway Maintenance Facility   Land developed and used for railway
                                               maintenance.
663   Railway Passenger Terminal Facilities Land developed and used as a railway passenger
            (including stations)   terminal.
664   Railway Freight Terminal Facilities   Land developed and used as a railway freight
                                               terminal.

665	Tramway/Light Rail Right of Way and Associated Track Infrastructure	Land developed and used as an operating tram or light rail service and associated infrastructure.
666	Tramway Maintenance /Terminal Storage	Land developed and used for tramway maintenance and terminal facilities.
668	Railway/Tramway Line Closed/ Unused	Land developed and no longer used for tramways or other related facilities.

67   *Transport – Air*

670	Airfield (includes associated open space)	Land developed with specialist infrastructure used as an airport capable of handling domestic and/or international services.
671	Airstrip	Land developed with limited infrastructure and used as a local/regional airstrip.
672	Airport Traffic Control Centre	Land developed with specialist infrastructure used and operated as an air traffic control centre.
673	Airport Hangar Building	Land developed with specialist infrastructure used for aircraft maintenance and storage.
674	Airport Terminal Building – Passengers	Land developed with specialist infrastructure used in conjunction with airline operations to manage passenger services.
675	Airport Terminal Building – Freight	Land developed with specialist infrastructure used for freight handling services within an airport.
676	Heliport	Land developed and used for helicopter landing and parking.

68   *Transport – Marine*

680	Port Channel	Designated channel used as a shipping waterway.
681	Port Dock/Berth	Seabed adjoining a wharf and developed with infrastructure used for the berthing of ships.
682	Port Wharf/Pier and Apron – Cargo	Land developed with specialist infrastructure to facilitate the movement of containers and cargo to and from ships.
683	Wharf – Storage Sheds	Land developed with enclosed storage facilities within a wharf.
684	Wharf – Passenger Terminal and Ferry Pier Facilities	Land developed with passenger terminals and other facilities within a wharf or pier.
685	Piers, Storages and Slipways	Land developed and used for the maintenance and launching of boats.
686	Ramps and Jetties	Land developed with limited infrastructure used for recreational boating purposes.
687	Marinas and Yacht Clubs	Land developed with specialist infrastructure used for the wet and dry storage of leisure boats.
688	Dockyard, Dry Dock and/or Ship Building Facility	Land developed with specialist infrastructure used for the repair, maintenance, and construction of ships.
689	Lighthouse and Navigation Aids	Land developed with specialist infrastructure used to assist in sea navigation.

Code	Infrastructure and utilities (industrial)	Description
69	**Communications, including Print, Post, Telecommunications and Airwave Facilities**	
690	Post Offices	Land used for the collection/ distribution of mail and the sale of products associated with that use.
691	Postal Exchange/Mail Sorting Centres	Land developed and used for the sorting of mail.
692	Post Boxes	Land developed with a single receptacle for the posting of mail.
693	Telecommunication Buildings/ Maintenance Depots	Land developed and used for the maintenance of telecommunication installations.
694	Telecommunication Towers and Aerials	Land developed with specialist infrastructure used for the transmission of telecommunication signals. Maybe standalone or affixed to buildings.
694.2	Telecommunication Tower	
694.3	Telecommunication Aerial	
695	Cable Lines, Conduits and Special Purpose Below Street Level Communication Line Tunnels – not being sewers (through easements, freehold and public land)	Land developed with cable lines, conduits and special purpose, below street level and communication line tunnels used for telecommunication purposes.
696	Television/Radio Station – Purpose Built	Land developed with specialist infrastructure used for the production/recording of television and radio programmes.
697	Printing Works/Press	Land developed with specialist infrastructure and used for printing works, e.g. newspaper print, magazines.
698	Telephone Exchange – Purpose Built	Land developed with specialist infrastructure used to facilitate the transmission of telephonics.

Code	Community Services	Description
70	**Vacant or Disused Community Services Site**	
700	Vacant Health Services Development Site	Vacant land with a permit approved or capable of being developed for health purposes, e.g. hospital site.
701	Vacant Education and Research Development Site	Vacant land with a permit approved or capable of being developed for education purposes, e.g. school/ university site.
702	Vacant Justice and Community Protection Development Site	Vacant land with a permit approved or capable of being developed for justice and community protection purposes, e.g. police station, court house.

Code	Community Services	Description
703	Vacant Religious Purposes Development Site	Vacant land with a permit approved or capable of being developed for religious purposes, e.g. church, temple, synagogue site.
704	Vacant Community Services Development Site	Vacant land with a permit approved or capable of being developed for community services, e.g. clubrooms.
705	Vacant Government Administration Development Site	Vacant land with a permit approved or capable of being developed for government administration purposes, e.g. civic purposes.
706	Vacant Defence Services Development Site	Vacant land with a permit approved or capable of being developed for defence purposes, e.g. barracks.
707	Cemetery	Land permitted to be used as a cemetery.

71 **Health**

710	Public Hospital	Land developed and used as a hospital funded by the government for public patients.
711	Private Hospital	Land developed and used as a non-government funded hospital for private patients.
712	Welfare Centre	Land developed and used for the purposes of providing welfare services to the community.
713	Community Health Centre	Land developed and used as consulting facilities, for a range of public health issues to the wider community.
714	Centre for the Mentally ill	Land developed with specialist facilities and used for the treatment of the mentally ill. Includes rehabilitation clinics.
715	Day Care Centre for Children	Land developed and permitted to be used as a day care centre of children by a licensed operator.

72 **Education and Research**

720	Early Childhood Development Centre – Kindergarten	Land developed and permitted to be used as a funded early education centre for children 3–5 year olds.
720.2	*Early Childhood Development Centre*	
720.3	*Kindergarten*	
720.4	*Pre-School*	
720.5	*Child Welfare and Pre-School*	
721	Government School	Land developed and used in the education of students in a government school operated by the state.
721.2	*Primary School*	Land developed and used in the education of students in years p-6/7.
721.3	*Secondary School/College*	Land developed and used in the education of students in years 7/8–12. Includes trade and technical schools.
721.4	*Combined Primary/Secondary*	Land developed and used in the education of students in years p-12.
721.5	*Technical School*	Land developed and used in the education of students in years 7/8–12 for trade and technical purposes.
721.6	*Playing Fields and Sporting Facilities*	
722	School Camps	Land developed and used as a camp exclusively for the education of students by a registered education provider.

723	Non-Government School	Land developed and used in the education of students in a non-government school operated by a registered education provider.
723.2	*Primary School*	Land developed and used in the education of students in years p-6/7.
723.3	*Secondary School/College*	Land developed and used in the education of students in years 7/8–12.
723.4	*Combined Primary/Secondary*	Land developed and used in the education of students in years p-12.
723.5	*Technical School*	Land developed and used in the education of students in years 7/8–12 for trade and technical purposes.
723.6	*Playing Fields and Sporting Facilities*	
724	Special Needs School	Land developed and used in the education of pupils with special needs by a registered education provider.
725	University	Land developed and used in undergraduate and post graduate studies at degree, masters and PhD levels by a registered university.
726	Technical and Further Education	Land developed and used for post-secondary school education and training by a registered education provider. Usually aimed at developing specific job core competencies.
727	Research Institute – Public	Land developed and used as a research facility by the government.
728	Observatory	Land developed with purpose-built infrastructure associated with astronomy and of national scientific importance.
729	Residential College/ Quarters Defence forces	Residential accommodation/quarters for the defence forces.
73	***Justice and Community Protection***	
730	Police Facility	Land developed and used as a policing facility, at district/ regional/state level.
731	Court Facility	Land developed and used as a judicial facility for either a court or tribunal.
732	Prison/Detention Centre/ Gaol Complex/ Corrective Institution	Land developed and used for custodial purposes.

Code	Community Services	Description
733	Fire Station Facility	Land developed and used for the storage of vehicles and equipment for the fighting of fires.
734	Ambulance Station Facility	Land developed and used as an ambulance station.
735	Emergency Services Complex	Land developed and used for state emergency services facilities.
736	Community Protection and Services Training Facility	Land developed and used as a specialist facility for the training of fire, police, ambulance, SES and prison personnel.

74	*Religious*	
740	Place of Worship	Land developed and used as a place of worship.
741	Religious Hall	Land developed and used for the social interaction of people by a religious organisation.
742	Religious Residence	Land developed and used as a dwelling by an ordained member/members of a religious order, as part of administering their religious duties.
743	Religious Study Centre	Land developed and used as a religious study centre.
75	**Community Service and Sporting Clubrooms and Halls**	
750	Halls and Service Clubrooms	Land developed and used as an occasional meeting place by community-based groups or clubs.
751	Rural and Community Camps	Land developed with accommodation used by persons or groups for short-term recreation/training/education purposes by a community service provider e.g. scout camp.
752	Community Facility	Land developed and used as a meeting place by groups involved in community interests, e.g. neighbour centre.
76	**Government Administration**	
760	Parliament House	Land developed and used by government as a house of parliament.
761	Government House	Land developed and used as a residence by a governor of a state or the Commonwealth.
762	Local Government	Land developed and used for the administration of local government.
763	Civic Buildings	Land developed and used by local government for civic purposes.
77	**Defence Services/Military Base**	
770	Army Barracks/ Administration Base	Land developed and used for the administration of the armed forces.
771	Army Maintenance Depots	Land developed and used by the army for the storage and maintenance of equipment and infrastructure.
772	Army Field Camps and Firing Ranges	Land developed and used by the army as field camps and for firearms training/practice.
773	Naval Base/Administration Base	Land developed and used for the administration of the naval forces.
774	Naval Specialised Facilities – Ground Based	Land developed with specialised infrastructure for use by the navy, e.g. storage, maintenance and training.
775	Naval Specialised Facilities – Water Based	Land developed with specialised infrastructure associated with the berthing of navy vessels, e.g. wharves, dry docks.
776	Air Force Base/ Administration	Land developed and used for the administration of the air forces.
777	Airstrip and Specialised Facilities	Land developed with specialised infrastructure associated with an airbase used by the Air Force, e.g. airfields, hangars, storage and maintenance.
778	Munitions Storage Facility	Land developed and used the defence forces for the storage of explosives, ammunition and bombs.

78	*Other Community service facilities*	
780	Public Conveniences	Land developed and used as a public convenience, e.g. public toilet block.
781	Unspecified – Public, Education and Health Improved	Land developed and used for the provision of education and health to the public by community service groups.
782	Unspecified – Public, Education and Health vacant	Vacant land with a permit approved or capable of being developed for public education and health services.
783	Animal shelter	Land developed and used as an animal welfare shelter by a community service group. May include keeping the animals on the premises for treatment or adoption
80	*Vacant Land*	
800	Vacant Site – Sporting Use	Vacant land with a permit approved or capable of being developed for sporting use.
801	Vacant Site – Heritage Application	Vacant land designated and zoned for heritage purposes, e.g. historic precinct, heritage landscape.
802	Vacant Site – Cultural Use	Vacant land designated and zoned for cultural purposes.
81	*State/Regional Sports Complex*	
810	Major Sports Complex	Land developed with specialised infrastructure used as a major sporting facility for commercial purposes, e.g. MCG.
811	Major Indoor Sports Complex	Land developed with specialised infrastructure used as an indoor sporting facility for commercial purposes, e.g. Rod Laver Arena.

Code	*Sport, Heritage and Culture*	*Description*
812	Outdoor\Indoor Sports Complex – non major	Land developed and used as a state or regional sports facility with limited commercial application, e.g. velodrome, netball, hockey centre.
813	Outdoor Sports – Extended Areas/Cross Country	Land developed with specialist infrastructure over extended open areas used for recreational/sporting activities, e.g. member facility golf course, polo fields.
814	Aquatic Complex	Land developed with specialised infrastructure used as an aquatic complex for water sports.
815	Water Sports – Outdoor	Land developed with specialised infrastructure used for open air water sports, e.g. rowing.
816	Motor Racing Tracks/ Speedways	Land developed with specialised infrastructure used for motor sports.
817	Racecourse/Tracks	Land developed with specialised infrastructure used for horse, greyhound, or harness racing, e.g. Flemington, Randwick.
818	Ski Fields	Land developed with specialised infrastructure used as a ski area for commercial purposes. Includes a field, run, trail or course prepared for the purpose of alpine recreation.

82	*Local Sporting Facilities*	
820	Indoor Sports Centre	Land developed and used as a local indoor recreational facility.
820.2	*Squash Courts*	
820.3	*Gymnasium/Health Club*	
820.4	*Indoor Sports Complex*	
820.5	*Bowling Alley*	
821	Outdoor Sports Grounds town or suburban facilities	Land developed and used as a local outdoor recreation facility.
821.2	*Tennis Club*	
821.3	*Bowling Club*	
821.4	*Outdoor Park and Facilities*	
822	Outdoor Sports – Extended Areas/Cross Country	Land developed with specialist infrastructure over extended open areas used for local recreational/ sporting activities, e.g. municipal golf course.
823	Swimming Pools/Aquatic Centres	Land developed with specialised infrastructure used as a local aquatic complex for water sports, e.g. municipal swimming centre.
824	Water Sports – Outdoor	Land developed with specialised infrastructure used for local open air water sports, e.g. rowing.
825	Motor Race Tracks/ Speedways	Land developed with specialised infrastructure used for local motor sports.
826	Aero Club Facility	Land used by aero clubs for flying pursuits. May include aircraft hangers.
827	Ski Fields	Land developed with limited infrastructure and used for recreational alpine pursuits. Limited commercial application.
828	Equestrian Centre	Land developed with specialised infrastructure used for the grooming and showing of horses.
829	Bike Track/Walking Trails	Land designated as a bike track/walking trail.
83	*National/State Cultural Heritage Centres*	
830	Library/Archives	Land developed and used as a library or archival facility with state or national significance.
831	Museum/Art Gallery	Land developed and used as a museum/art gallery with state or national significance.
832	Cultural Heritage Centre	Land developed and used as a cultural heritage centre with state or national significance.
833	Wildlife Zoo	Land developed and used as zoological gardens with state or national significance.
834	Aquarium	Land developed and used as an aquarium with state or national significance.
835	Botanical Gardens	Land developed and used as botanical gardens with state or national significance.
836	Monument/Memorial	Land developed and used as a monument/memorial with state or national significance, e.g. Shrine of Remembrance.
837	Culture, recreation and sport	Land developed and used as a culture, recreation and sport centre with state or national significance.

84	*Local Cultural Heritage Sites, Memorials and Monuments*	
840	Library/Archives	Land developed and used as a library or archival facility with local significance.
841	Museum/Art Gallery	Land developed and used as a museum/art gallery with local significance.
842	Cultural Heritage Centre	Land developed and used as a cultural heritage centre with local significance.
843	Wildlife Zoo/Park/ Aquarium	Land developed and used as a wildlife zoo/aquarium with local significance.
844	Parks and Gardens	Land developed and used as parks and gardens with local significance.
845	Monument/Memorial	Land developed and used as a monument/memorial with local significance.
85	*Local Recreation – Other*	
850	Bathing Boxes	Land developed on the foreshore and used as a bathing box for recreational purposes.
851	Boat Sheds	Land developed on the foreshore and used as a boat shed for recreational purposes.

Code	National parks, conservation areas, forest reserves and natural water reserves	Description
90	*Reserved Land*	
900	Vacant Land	Vacant land with special conservation values designated but not proclaimed as a reserve.
91	*Nature Reserve*	
910	Nature Reserve	Land designated and proclaimed as a nature reserve.
92	*Wilderness Area*	
920	World Heritage Area	Land designated and proclaimed as a world heritage area. Recognised internationally for its unique wilderness values.
921	Local Wilderness Area	Land designated and recognised as a local wilderness area.
93	*National Park (Land and Marine)*	
930	National Park – Land	Land designated and proclaimed as a national park.
931	National Park – Marine	Land designated and proclaimed as a national marine park.
94	*Natural Monument/Feature*	
940	Natural Monument – Land	Land recognised for its renowned features/scenic/natural/ cultural values, e.g. Three Sisters.
941	Natural Monument – Marine	Land recognised for its marine features/scenic/cultural values, e.g. Twelve Apostles – Victoria.

95 *Natural Forests and Forest Reserves*

| 950 | Forest Reserves – Public | Public land reserved for the preservation or protection of aesthetic, scientific, flora or fauna values. |
| 951 | Forest Reserves – Private | Private land reserved for the preservation or protection of aesthetic, scientific, flora or fauna values. |

96 *Conservation Area*

| 960 | Conservation Area – Public | Public land predominantly in a natural state designated as a conservation area. |
| 961 | Conservation Area – Private | Private land predominantly in a natural state designated as a conservation area. |

97 *Protected Landscape/ Seascape*

970	Protected Landscape – Public	Public land designated as a protected landscape recognised for its natural and cultural values.
971	Protected Landscape – Private	Private land designated as a protected landscape recognised for its natural and cultural values.
972	Protected Seascape – Public	Public land designated as a protected seascape recognised for its natural and cultural values.
973	Protected Seascape – Private	Private land designated as a protected seascape recognised for its natural and cultural values.

98 *Wetlands*

980	River Reserve (fresh water)	Land designated as a freshwater river reserve, usually with all year round flows.
981	Creek Reserve (fresh water)	Land designated as a fresh water creek reserve with intermittent flows and tides.
982	River Reserve (salt water)	Land designated as a salt water river reserve, usually with all year round flows.
983	Creek Reserve (salt water)	Land designated as a salt water creek reserve with intermittent flows and tides.
984	Floodway Reserve	Land designated as a floodway reserve.
985	Fresh Water Lake Reserve	Land designated as a fresh water lake reserve that usually holds water all year round.
986	Salt Water Lake Reserve	Land designated as a salt water lake reserve and is not used for commercial salt extraction.
987	Inland Low Lying Tidal Estuary Wetlands Reserve	Land designated as a wetlands reserve associated with enclosed bays/salt water river estuary.
988	Seabed – Open Sea/Ocean/ Bays	Open sea below high water mark, not being a marine park.

99 *Game/Fauna Reserves*

| 990 | Game Reserve – Public | Public land designated as a game reserve. Hunting of game may be permitted. |
| 991 | Game Reserve – Private | Private land designated as a game reserve. Hunting of game may be permitted. |

INDEX